ION EXCHANGE
Theory and Practice

Second Edition

Royal Society of Chemistry Paperbacks

Royal Society of Chemistry Paperbacks are a series of inexpensive texts suitable for teachers and students and give a clear, readable introduction to selected topics in chemistry. They should also appeal to the general chemist. For further information on selected titles contact:

Sales and Promotion Department
The Royal Society of Chemistry
Thomas Graham House
The Science Park
Milton Road
Cambridge CB4 4WF, UK

Titles Available

Water *by Felix Franks*
Analysis – What Analytical Chemists Do *by Julian Tyson*
Basic Principles of Colloid Science *by D. H. Everett*
Food – The Chemistry of Its Components (Second Edition)
by T. P. Coultate
The Chemistry of Polymers *by J. W. Nicholson*
Vitamin C – Its Chemistry and Biochemistry
by M. B. Davies, J. Austin, and D. A. Partridge
The Chemistry and Physics of Coatings
edited by A. R. Marrion
Ion Exchange: Theory and Practice, Second Edition
by C. E. Harland

How to Obtain RSC Paperbacks

Existing titles may be obtained from the address below. Future titles may be obtained immediately on publication by placing a standing order for RSC Paperbacks. All orders should be addressed to:

The Royal Society of Chemistry
Turpin Distribution Services Limited
Blackhorse Road
Letchworth
Herts SG6 1HN, UK

Telephone: +44 (0) 462 672555
Fax: +44 (0) 462 480947

Royal Society of Chemistry Paperbacks

ION EXCHANGE
Theory and Practice

Second Edition

C. E. HARLAND

*The Permutit Company Limited
(Part of Thames Water plc), UK*

ISBN 0-85186-484-8

A catalogue record for this book is available from the British Library

© The Royal Society of Chemistry, 1994

All Rights Reserved
No part of this book may be reproduced or transmitted in any form or by any means – graphic, electronic, including photocopying, recording, taping, or information storage and retrieval systems – without written permission from the Royal Society of Chemistry

Published by The Royal Society of Chemistry,
Thomas Graham House, The Science Park,
Milton Road, Cambridge CB4 4WF, UK

Typeset by Keytec Typesetting Ltd, Bridport, Dorset
Printed in Great Britain by The Bath Press, Lower Bristol Road, Bath

Preface

AIMS

This publication is a revision of the first edition issued in 1975, as part of the then Chemical Society's highly successful 'Monographs for Teachers' series. Many significant advances have been made during the intervening years with regard to process development, ion exchange materials, equipment engineering, and the continuing quest for a better understanding of fundamental principles.

The 1990s see the realization of over fifty years growth in the principal application of ion exchange; therefore the aims and objectives of this revision are deliberately somewhat different from the earlier edition. The salient material concerning the early history of the subject is retained, but in dealing with modern developments more emphasis is placed on the properties of modern resins and the interrelationship between processes and fundamental theory. In this way it is hoped to sacrifice a formal catalogued review approach in favour of cultivating interest and better understanding. Ion exchange using modern exchangers is an excellent vehicle for experimentation and demonstration of not only key topics in a school's core curriculum science syllabus, but also as a possible component of student project investigations into any one of a host of electrolyte solution chemistry topics. In order to support an interest in the latter, simplified procedures for characterizing organic exchangers and selected bench experiments have been included in separately identified text (**Boxes**), but it is also hoped that, within teaching institutions, students will be encouraged to recognize the potential use of ion exchangers in their own experimental and project programmes (see Note 1).

In industry, not all personnel responsible for either purchasing or operating ion exchange plant have necessarily been formally versed in the mysteries of the subject. With this fact in mind this paperback, within the constraints imposed by length, sets out to explain the principal elements concerning exchanger characteristics, fundamental theory, and key process applications. It remains the prerogative of reviewers and readers to decide how well the author's stated aims have been met to produce an informative paperback which complements the more rigorous and comprehensive texts currently available.

An exhaustive literature review is purposely not included, but instead, principal subject categories are referred to a bibliography by chapter listing authoritative texts and publications by which means a more detailed study of a topic and access to the formal scientific literature may be accomplished.

THE SUBJECT

Observations of the phenomenon of ion exchange date from ancient times. The 'mechanism' of the reaction, however, was established in 1850 by two English chemists, H. S. Thompson and J. T. Way, but it is only in the past few decades that the subject has expanded to become a true science from which extensive industrial applications have emerged. Progress in both the theoretical understanding and process developments of ion exchange has been rapid with the one complementing the other. In the academic sphere, the growth of knowledge has paralleled closely the physical chemistry interpretation of the behaviour of concentrated solutions of electrolytes. This is not unexpected since ion exchange materials are in effect electrolytes, albeit mostly solid ones, and in some respects they behave in an identical fashion. Undoubtedly, present day inadequacies in the theory and understanding of certain aspects of ion exchange stem from the lack of knowledge and precise interpretation of the various forces and effects of electrical origin which govern the chemical behaviour of electrolyte solutions.

The scale and scope of industrial ion exchange has grown enormously since the early days of water softening, through which the subject achieved worldwide acclaim. Water treatment is still the largest single industrial application of ion exchange, and at present there is a large and increasing demand not only for softened water, but also for 'pure' or demineralized water. Vast amounts of such high quality water are essential to many highly technological industries such as those producing fabricated metals, paper, synthetic fibres, electronic components, processed foodstuffs, pharmaceuticals, and electrical power. Much current research and process development is directed towards utilizing ion exchange methods for obtaining potable water from sources which in one way or another fail to meet current European Union (EU – formerly European Community or EC) directives, for example high nitrate levels or excessive salinity.

'Matter may be neither created nor destroyed' but, if no one is looking, it may be thrown away. However, people are looking and the public awareness of a need to control effluent disposal, and so reduce environmental pollution, has never been higher. The practice of ion exchange in the field of effluent treatment is well established, and its potential uses in these areas with the added advantage of reagent recovery and water re-use are always under active review.

In many mineral fields it is becoming increasingly apparent that, although the demand for growth increases, the natural resources and quality of raw materials are decreasing. This is particularly true in many aspects of extraction metallurgy where good quality ore bodies are being rapidly depleted and man is being compelled to utilize lower grade materials which frequently demand new and more efficient processing techniques. It is in many such applications that ion exchange has proved extremely useful and it is to be expected that the techniques will be used on an ever-increasing scale in the future.

In medicine, also, considerable use is being made of ion exchangers, particularly in the production of water to meet the stringent quality specification required for pharmaceutical and cosmetic formulations. The controlled, slow release of drugs and other chemicals into the body has also been made possible by means of these versatile components.

The earliest references to ion exchange are in relation to soils and fertility. It may not be surprising, therefore, that modern synthetic exchangers have wide potential application in agriculture and horticulture. Elements vital to plant growth may be introduced to soils and other fertile media by means of ion exchangers, from which they may be liberated at a controllable rate.

Finally, any preface to the subject of ion exchange cannot omit to highlight the impact made by ion exchange chromatography; not only historically through its firstly being responsible for the separation of chemically similar species such as the lanthanides, actinides, and amino acids, but also the more recent achievement whereby microquantities of ion mixtures may be rapidly separated and individual components quantitatively detected at sub-microgram levels.

Note 1

BOXES describing experiments and demonstrations may be read separately from the main text, and the following identities apply where appropriate:

ml	≡	cm^3
%, or %w/v	≡	grams per 100 cm^3 solvent
normality, N	≡	g equiv. l^{-1}
BV	≡	'tapped down' bed volume
SAC	≡	strong acid cation resin
WAC	≡	weak acid cation resin
SBA	≡	strong base anion resin
WBA	≡	weak base anion resin

Where an estimated time is stated it is assumed that standard reagents, resins in the appropriate ionic form, and all other equipment items are immediately available. Where deionized water is specified, distilled water may be used as an alternative. Where methyl orange indicator is stated, BDH 4.5 indicator may be substituted if preferred.

Contents

Preface v
Aims v
The Subject vi

Boxes xiii

Acknowledgements xv

Chapter 1
Discovery and Structure of Solid Inorganic Ion Exchange Materials 1
The Phenomenon 1
Inorganic Materials 2
Ion Exchange Properties of the Aluminosilicates 10
Further Reading 19

Chapter 2
The Development of Organic Ion Exchange Resins 21
Early Organic Ion Exchange Materials 22
Modern Organic Ion Exchange Resins 25
Special Ion Exchange Materials 32
Further Reading 37

Chapter 3
Structure of Ion Exchange Resins — 39

Conventional Resin Structure	39
Solvent Modified Resin Structures	45
Acrylic Anion Exchange Resins	46
Solvent Modified Resin Adsorbents	47
Further Reading	48

Chapter 4
Properties and Characterization of Ion Exchange Resins — 49

Resin Description	49
Chemical Specification	58
Physical Specification	82
Summary	87
Further Reading	88

Chapter 5
Ion Exchange Equilibria — 90

Introduction	90
Swelling Phenomena and the Sorption of Solvents	93
Sorption of Non-exchange Electrolyte and the Donnan Equilibrium	101
Relative Affinity	104
Selectivity Coefficient	105
Rational Thermodynamic Selectivity	111
Prediction and Interpretation of Selectivity	113
Dilute Solution Cation Exchange	123
Dilute Solution Anion Exchange	127
Selectivity in Concentrated Solutions	131
Further Reading	132

Chapter 6
The Kinetics and Mechanism of Ion Exchange — 134

Basic Concepts	134
Rate Equations	140
Mechanism Criteria	154

Column Dynamics	158
Breakthrough Curve Profiles	161
Process Design	164
Further Reading	164

Chapter 7
Some Basic Principles of Industrial Practice — 166

Introduction	166
Column Operations	167
Operating Capacity and Regeneration Efficiency	169
Column Breakthrough and 'Leakage'	173

Chapter 8
Water Treatment — 179

Water Analysis	180
Softening	187
Dealkalization	192
Other Single Cycle Ion Exchange Processes in Water Treatment	197
Demineralization	204
Coflow Two-stage Systems	205
Coflow Multistage Processes	212
Counterflow Systems	213
Combined Cycle Single Stage Demineralization	215
Desalination by Ion Exchange	226
Waste Effluent Treatment by Ion Exchange	228
Further Reading	236

Chapter 9
Non-water Treatment Practices — 238

Carbohydrate Refining	238
Catalysis	241
Metathesis	244
Recovery Processes	246
Pharmaceutical Processing	252
Ion Separation	253
Further Reading	259

Chapter 10
Some Engineering Notes 261

Conventional Plant	261
Continuous Countercurrent Ion Exchange	270
The Past, Present, and Future	273
Further Reading	276

Appendix 277

Useful Conversions	277
Some Standard Resins According to Generic Type and Manufacturer	278

Subject Index 280

Boxes

1.1	Demonstration of the Nature of Inter-lamellar Bonding within Double Layer Aluminosilicates	14
3.1	Demonstration of Structural Strain during Resin Swelling	43
4.1	Experiment to Demonstrate the Dissociation Properties of Strong Functional Groups	51
4.2	Experiment to Illustrate the Difference in Properties of Weak and Strong Functional Groups	56
4.3	Conversion of Resins to Standard or Alternative Ionic Forms	63
4.4	Simplified Measurement of Water Regain and Voids Volume	66
4.5	Determination of Dry Weight Capacity	73
4.6	Determination of Wet Volume Cation Capacity	76
4.7	Determination of Strong and Weak Base Wet Volume Capacity of an Anion Exchange Resin	78
4.8	Measurement of Swollen Resin Density	85
5.1	A Simplified Experiment to Demonstrate Affinity Sequences for Ion Exchange Reactions and Estimation of the Selectivity Coefficient	108
5.2	Sorption of a Cation as an Anionic Complex	131

6.1	An Experiment to Show 'Moving Boundary' Formation during Particle Diffusion Accompanied by Chemical Reaction	139
6.2	A Simple Rate Experiment	153
8.1	The Determination of Total Anions and Cations in a Solution	185
8.2	Softening or Displacement of a Monovalent Ion by a Divalent Ion	189
8.3	Two-stage Demineralization of a Solution	211
9.1	Decolorizing and Demineralizing of a Sugar Solution	240
9.2	Catalysis – Hydrolysis of Ethyl Acetate by Means of a Cation Resin	242
9.3	Metathesis Reactions	245
9.4	Concentration of a Metal from a Dilute Solution	250
9.5	Elution Chromatography of Simple Cations	254

Acknowledgements

I would like to gratefully acknowledge the valuable assistance and encouragement given by family, friends, and colleagues towards the production of this book.

Firstly I would like to thank my family, and especially my wife Helen, for their support and encouragement.

I thank my friends and colleagues within The Permutit Company Ltd and PWT Projects for their technical assistance and patience whilst proofreading, and the Permutit Company Ltd for their permission to publish.

A special appreciation is extended to Denise Pearsall for her enthusiasm, devotion to the task, and patience whilst typing the draft copy. I thank also Mr Norman Robertson for his computer drafting of tabulated material.

C. E. Harland
October 1993

Chapter 1

Discovery and Structure of Solid Inorganic Ion Exchange Materials

THE PHENOMENON

An ion exchange reaction may be defined as the reversible interchange of ions between a solid phase (the ion exchanger) and a solution phase, the ion exchanger being insoluble in the medium in which the exchange is carried out. If an ion exchanger M^-A^+, carrying cations A^+ as the exchanger ions, is placed in an aqueous solution phase containing B^+ cations, an ion exchange reaction takes place which may be represented by the following equation:*

$$\underset{\text{Solid}}{M^-A^+} + \underset{\text{Solution}}{B^+} \rightleftharpoons \underset{\text{Solid}}{M^-B^+} + \underset{\text{Solution}}{A^+} \quad (1.1)$$

For reasons which will be considered later, the anion in solution does not necessarily take part in the exchange to any appreciable extent. The equilibrium represented by the above equation is an example of *cation exchange*, where M^- is the insoluble fixed anionic complement of the ion exchanger M^-A^+, often called simply the *fixed anion*. The cations A^+ and B^+ are referred to as *counter-ions*, whilst ions in the solution which bear the same charge as the fixed anion of the exchanger are called *co-ions*. In much the same way, anions can be exchanged provided that an anion-receptive medium is employed. An analogous representation of an *anion exchange* reaction may be written:

$$\underset{\text{Solid}}{M^+A^-} + \underset{\text{Solution}}{B^-} \rightleftharpoons \underset{\text{Solid}}{M^+B^-} + \underset{\text{Solution}}{A^-} \quad (1.2)$$

---

*An ion may be defined as an atom or combination of atoms (molecule) which carry a net positive (cation) or net negative (anion) electrical charge.

Further development of a physical model for the exchanger phase is best left until Chapter 2 when synthetic ion exchangers will be considered in more detail, but the previous equations illustrate the essential difference between ion exchange and other sorption phenomena. The main fact is that electroneutrality is preserved at all times in both the exchanger and solution phases, and this in turn requires that counter-ions are exchanged in *equivalent* amounts. The most important features characterizing an ideal exchanger are:

1. A hydrophilic structure of regular and reproducible form.
2. Controlled and effective ion exchange capacity.
3. Rapid rate of exchange.
4. Chemical stability.
5. Physical stability in terms of mechanical strength and resistance to attrition.
6. Consistent particle size and effective surface area compatible with the hydraulic design requirements for large scale plant.

Manufacturers of modern ion exchange materials have progressed a long way towards meeting all these requirements when compared with the prototype materials described below. The cost to industry of modern ion exchange resins is high, varying typically from £1000 to £4000 per m^3. Therefore exchanger properties which minimize the volumes required (*e.g.* high exchange capacity), or which prolong resin life (*e.g.* physical and chemical stability), are important considerations. It therefore follows that continued efforts to improve the exchanger characteristics listed above play an important part in the activities of resin manufacturing companies.

INORGANIC MATERIALS

References to ion exchange phenomena have been attributed to Old Testament scribes, and later to Aristotle, but the first descriptions in modern scientific terminology have been credited to two English soil chemists, H. S. Thompson and J. T. Way in the mid-nineteenth century. They observed 'base' or cation exchange between calcium and ammonium ions on some types of soil. Upon treating a column of soil with a solution of ammonium sulfate it was found that most of the ammonia was absorbed, whilst the calcium contained originally in the soil was released and passed out of the column. Further studies furnished many sound conclusions as to the nature of ion exchange reactions, some of the more important ones being:

1. Exchange involved equivalent quantities of ions.
2. Certain ions were more easily exchanged than others.
3. The temperature coefficient for the rate of exchange was small.
4. The aluminosilicate fractions of soils were responsible in the main for the exchange although these components rarely took part in the exchange itself.*
5. Materials possessing exchange properties could be synthesized from soluble silicates and aluminium sulfate.

The equivalence law governing the phenomenon was established in the early scientific history of the subject, as also was the fact that some ions were more easily exchanged than others; in other words ion exchangers showed greater selectivity or affinity for different ions. That an exchanger could be chemically synthesized proved to be of the utmost importance; it is for this reason that ion exchange studies and applications have reached such an advanced state today.

The ion exchange capacity of an exchanger is a measure of its total content of exchangeable ions, and is conventionally expressed in terms of the total number of equivalents of ion per kilogram (milli-equivalents per gram) of the exchanger in its dry state and in a given univalent ionic form. As will become evident when describing practical applications, the operating exchange capacity of an exchanger is invariably less than its total capacity. Also because of the presence of 'colloidal humus' in natural soils, exchange capacity data was difficult to systematize and to reproduce in relation to the inorganic minerals which were present. Consequently further studies were carried out utilizing the separated microcrystalline aluminosilicates, or 'clay fractions' of the soil which were obtainable in quite a pure form. These experiments proved that the main exchange agents were indeed contained in the finest or clay-like fractions of soils. Our knowledge of the structure and classification of such materials has shown that some of the inherently finely divided clay materials are directly responsible for the exchange characteristics observed; however, the phenomenon is not purely a property of particle size.

Why the clay minerals should possess an appreciable exchange capacity became more fully understood with the establishment of the crystal structures of the various types, for which most of the early credit is due to W. L. Bragg, L. Pauling, and others. Therefore before

---

* In all soils there is present a fraction called 'colloidal humus' which also contributes to the exchange capacity. This comprises very large, high molecular mass compounds containing organic amino, hydroxy, and carboxylic acid groups originating from vegetable decay, which are often bound with silica and heavy metals such as iron.

the ion exchange relationships of such materials can be fully understood a general appreciation of their structures is essential.

Geologically and genetically, clay minerals are difficult to define simply and adequately, but broadly they are layer lattice silicates of secondary origin. In the same classification are the micas, talc, chlorites, and serpentines which are not strictly clay minerals. In this context, secondary origin means that mineral formation has arisen from the weathering of primary or igneous rock, *e.g.* granites and basalts.

The basic structural unit making up the layer lattice silicates is the silica tetrahedron, $(SiO_4)^{4-}$. When three oxygen atoms of every tetrahedron are linked to similar units a continuous sheet structure is formed which is capable of indefinite extension in two directions at right angles; as a consequence the important physical property of minerals within this group is their plate-like character. One oxygen atom of each tetrahedral unit is not satisfied electrically and requires to be linked to external cations in order to establish electrical neutrality within the lattice. In most structures of this type the silica units are arranged in the form of hexagonal rings, each of which is surrounded by six similar ones, so that bonding takes place by the silica tetrahedra sharing three corners, as shown in Figure 1.1.

Each silica unit in the hexagonal sheet is linked to others through an oxygen atom. The Si—O—Si bond angle can vary, thus giving rise to different conformations of the ring structure, but the majority are based on a bond angle of about 141° 34′. As a result, the single oxygen atoms of each silica tetrahedron which are unsatisfied electrically are oriented in the same spatial direction.

Electrical neutrality in layer lattice silicates is maintained by condensing a sheet hydroxide structure with the sheet of silica tetrahedra (see Figure 1.2). The two types of structure which can take part in combinations of this nature are the hydroxides of divalent elements such as magnesium and those of trivalent elements such as aluminium. In both cases the cations are in six-fold co-ordination with anionic units, but whereas these are entirely hydroxide ions in the case of the pure hydroxide forms (namely brucite and gibbsite respectively) in the layer lattice silicates both hydroxide and oxygen ions are involved. The resulting layer lattice is theoretically electrically balanced within itself, and although the structure is capable of indefinite extension in two dimensions by ionic/covalent bonds, no similar continuity is possible in the direction at right angles to the basal plane. A further condensation of a silica layer in an inverted form above the hydroxide layer can also occur, thus increasing the

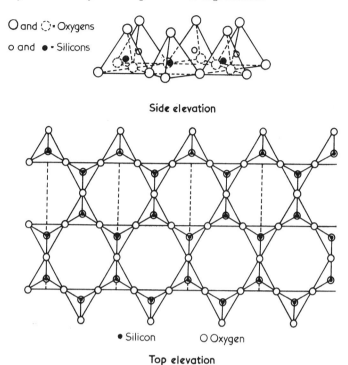

Figure 1.1 *Structure of the ideal silica layer of layer lattice silicates* (Reproduced by permission from R. W. Grimshaw, 'The Chemistry and Physics of Clays', E. Benn, London, 1971)

size and complexity of the layer and giving rise to other forms of layer lattice silicate minerals.

There are two major subdivisions of layer lattice silicates: a *single layer* type based on a condensation of a hydroxide layer structure with one silica plane and a *double layer* unit in which a further inverted silica plane completes a sandwich-like structure above the hydroxide unit. Each layer lattice is theoretically complete within itself and although similar layers can stack above each other there can be no formal inter-layer ionic or covalent bond formation.

Single Layer Lattice Silicates

Trivalent cations. By far the most common mineral group of this category is that of the kaolin minerals or Kandite group. The cations involved are solely aluminium which are each linked to three hydroxyl units in one layer and to two oxygens and one hydroxyl in the other.

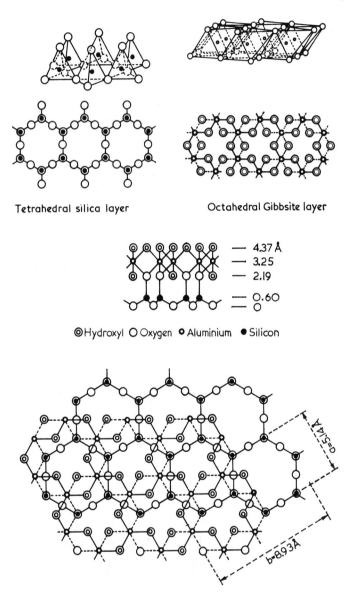

Figure 1.2 *Condensation of silica and gibbsite layers to give the kaolin layer structure* (nm = Å × 10⁻¹)
(Reproduced by permission from R. W. Grimshaw, 'The Chemistry and Physics of Clays', E. Benn, London, 1971)

The structure is illustrated in Figure 1.3. The kaolin layer is electrically neutral but extension in the c-crystallagraphic direction is possible through hydrogen bonding. This weak linkage results in a plate-like or flaky crystal habit.

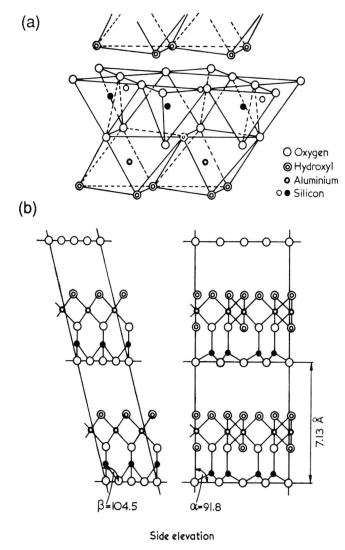

Figure 1.3 *Diagrammatic structures of the kaolin layer*
(Reproduced by permission from (a) R. E. Grim, 'Clay Mineralogy', McGraw–Hill, London, 1953, and (b) R. W. Grimshaw, 'The Chemistry and Physics of Clays', E. Benn, London, 1971)

Because several spatial stacking arrangements are possible there are several kaolin minerals, each with the same chemical composition, namely $Al_2Si_2O_5(OH)_4$, but with different properties. Nacrite, dickite, kaolinite, halloysite, and livesite are well recognized species. No positive evidence has so far been published linking other trivalent cations with a single layer lattice structure, but it has been suggested that iron(III) can replace aluminium in part in the kaolin lattice.

Divalent Cations. Both magnesium and iron(II) can take part in single layer lattice silicate structures although the former is more common. The cations are also in octahedral co-ordination but in order to preserve electrical balance within the lattice, all three of the octahedral sites over each silica hexagonal ring are occupied, as against two in the equivalent kaolin lattice. Divalent cations thus form a *trioctahedral* series whereas the trivalent cation minerals are termed a *dioctahedral* series. The typical magnesium structure is antigorite $Mg_3Si_2O_5(OH)_4$, but there are other minerals of this group which probably differ by virtue of the layer stacking isomerism and in the degree of substitution of ion(II). Chrysotile, the common asbestos mineral, chamosites, and some chlorites are typical examples.

Double Layer Lattice Silicates

As in the single layer group, two characteristic types of unit are to be found in the double layer silicate minerals based on the valency of the counterbalancing cations. A dioctahedral series, based on pyrophyllite $Al_2(Si_2O_5)_2(OH)_2$ and a trioctahedral group with talc $Mg_3(Si_2O_5)_2(OH)_2$ as the type minerals are well established. These structures are depicted in Figure 1.4. In marked contrast with minerals of the previous group, extensive elemental substitution within the double layer lattice is the rule rather than the exception. Not only does replacement of ions of identical charge and similar size occur but more complex substitution giving rise to charge deficiencies is quite common.

In *micas*, for example, aluminium ions replace silicon in the outermost tetrahedral layers of the lattice thus giving rise to a unit charge deficiency for every replacement. Electrical neutrality is then achieved by incorporating alkali or alkaline earth cations between the individual structural layers. Some typical structural formulae are:

$KAl_2(AlSi_3)O_{10}(OH)_2$ – dioctahedral muscovite mica
$K(Mg, Fe)_3(AlSi_3)O_{10}(OH)_2$ – trioctahedral biotite mica

Discovery and Structure of Solid Inorganic Ion Exchange Materials

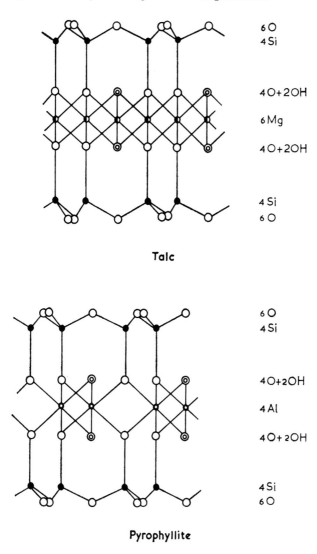

Figure 1.4 *Side elevation structures of talc and pyrophyllite*
(Reproduced by permission from R. W. Grimshaw, 'The Chemistry and Physics of Clays', E. Benn, London, 1971)

Montmorillonite minerals or smectites are based on a layer lattice in which the ionic replacement is mainly in the central octahedral layer. Once again the ionic substitution introduces a charge deficiency, typically Mg^{2+} for Al^{3+}. Counterbalancing hydrated cations occupy inter-layer positions, but because the charge deficiencies are situated in the centre of the layer, and are generally smaller in amount, the binding forces are less rigid than those in micas. The extra ions are thus less firmly held and are readily exchanged. Many montmorillonite minerals are known of both dioctahedral and trioctahedral types. Typical montmorillonite and mica structures are shown in Figures 1.5 and 1.6. Other minerals with a double layer structure are chlorites and vermiculites, where lattice charge deficiencies are counterbalanced by either hydroxide layers or hydrated cations between the layers.

All the clay minerals are of relatively small crystallite size, possessing a high specific surface area with many broken edge and surface bonds. In addition, the double layer group minerals frequently contain unbalanced electrical forces within the lattice, all of which have an influence on the overall electrical field surrounding the particles when they are suspended in a polar liquid medium. It is just such complex surface charge properties that play an important role in determining our understanding of such processes as the clarification and filtration of natural water supplies using coagulants such as aluminium sulfate.

ION EXCHANGE PROPERTIES OF THE ALUMINOSILICATES

Single Layer Lattice Silicates

The ideal constitution of the kaolin layer represents an electrically neutral unit, with rarely any isomorphous substitution of cations of different charges within the lattice. Consequently, kaolinite and related minerals would not be expected to show a large cation exchange capacity, and indeed this is usually the case. That a small but varying exchange capacity does occur may be attributed to two principal causes.

1. *Broken Bonds.* The naturally occurring single layer lattice clay minerals do not constitute a perfectly crystalline state, and around the edges of the silica–alumina layers, broken bonds give rise to unsatisfied negative charges which may be balanced by adsorbed cations. The size of the clay mineral particles and the number of lattice distortions play a part in determining the extent of ion exchange capacity, but it remains a matter of some

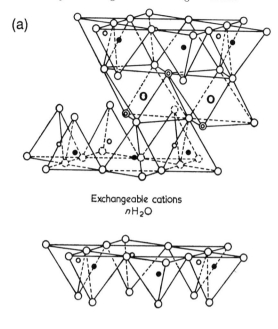

Exchangeable cations
$n\text{H}_2\text{O}$

O Oxygen ⊚ Hydroxyl ◯ Aluminium, iron, magnesium
◦ and ● Silicon, occasionally aluminium

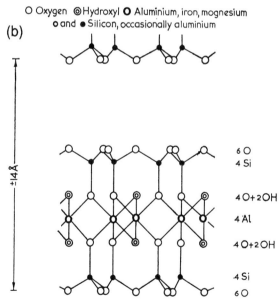

Figure 1.5 *Diagrammatic structure of montmorillonite*
(Reproduced by permission from (a) R. E. Grim, 'Clay Mineralogy', McGraw–Hill, London, 1953, and (b) R. W. Grimshaw, 'The Chemistry and Physics of Clays', E. Benn, London, 1971)

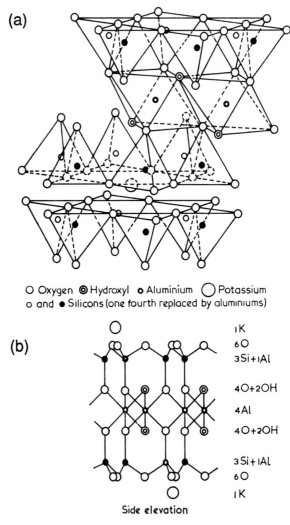

○ Oxygen ◉ Hydroxyl ○ Aluminium ○ Potassium
○ and ● Silicons (one fourth replaced by aluminiums)

Side elevation

Figure 1.6 *Diagrammatic structure of muscovite mica*
(Reproduced by permission from (a) R. E. Grim, 'Clay Mineralogy', McGraw-Hill, London, 1953, and (b) R. W. Grimshaw, 'The Chemistry and Physics of Clays', E. Benn, London, 1971)

conjecture whether true cation exchange is manifested by extensive breakdown of the crystal lattice (Tables 1.1 and 1.2).

2. *The hydrogen of exposed hydroxyls.* A further contribution to the exchange capacity of clay minerals is made by the hydrogens of

Table 1.1 *Variations in the cation exchange capacity of kaolinite with particle size* (from C. G. Harmon and F. Fraulini, *J. Am. Ceram. Soc.*, 1940, **23**, 252)

Particle size (μm)	Exchange capacity (eq kg^{-1})
10–20	0.024
5–10	0.026
2–4	0.036
0.5–1	0.038
0.25–0.5	0.039
0.1–0.25	0.054
0.05–0.1	0.095

Table 1.2 *Cation exchange capacity of kaolinite in relation to the time of grinding* (from W. P. Kelly and H. Jenny, *Soil Sci.*, 1936, **41**, 367)

Mineral	Exchange Capacity (eq kg^{-1})
Kaolinite (152 μm)	0.08
Kaolinite ground 48 hours	0.58
Kaolinite ground 72 hours	0.70
Kaolinite ground 7 days	1.01

exposed hydroxyl groups which may be replaced by exchangeable cations. This cause of exchange capacity is of particular significance for clay minerals where there exists an exposed sheet of hydroxyls on one side of the basal cleavage plane. Exposed hydroxyl groups may also exhibit a slight but reversible affinity for exchangeable anions. Isomorphous substitution in the kaolin lattice may occur, but it is not a major contributory factor to the ion exchange capacity unlike the case with some double layer lattice silicates.

Double Layer Lattice Silicates

Some of these minerals possess high cation exchange capacities greatly in excess of that attributable to surface area, crystal fracture and edge effects. The reason for this difference in behaviour arises out of a third and major cause of exchange capacity, namely isomorphous substitution resulting in charge deficiencies in the layer lattice. The montmorillonites typify such minerals of this type. The description of these minerals has been given earlier when it was stated that the counterbalancing cations are held rather loosely between the layer planes (see Figure 1.5). Because isomorphous substitution occurs in the octahedral layer the resulting positive charge deficiency is relatively delocalized with respect to the inter-lamellar plane. Although the cations may effect some kind of inter-layer bonding they are nevertheless readily exchanged by means of simple reversible diffusion between external solution and inter-lamellar sites.

As a consequence of this weak inter-layer bonding, the montmorillonite-type minerals can expand reversibly during the sorption of liquids and solvated ions, and are often termed expanding layer lattice silicates. When natural flakes of this family of materials are heated, for example vermiculite, the material expands irreversibly with startling visual effect to take the form of the familiar puffy granules, marketed as 'exfoliated vermiculite' and used as a horticultural growing medium and an insulating material (Box 1.1). Normally the basal spacing is about 1.4 nm (14 Å) but this value varies greatly with the nature of the ions and solvent being sorbed. The nature of solvents, especially water, held in the inter-layer spaces in montmorillonites is still not fully understood.

> **BOX 1.1 Demonstration of the Nature of Inter-lamellar Bonding within Double Layer Aluminosilicates**
>
> The difference in strength of inter-layer bonding between expanding and non-expanding double layer lattice silicates may be easily observed by the following demonstration:
> Gently heat flakes of natural vermiculite spread on a gauze over a bunsen flame. The material rapidly expands (exfoliates) to give puffy granules. Carry out the same test on flakes of mica whereupon nothing happens showing that the mica lattice is non-expanding and resistant to heat.
> **Caution**: Safety goggles should be worn.

To state categorically that isomorphous substitution in the montmorillonites concerns only the octahedral layer would be misleading. Some tetrahedral replacement could and does occur, but for most minerals within this class the observed exchange capacity correlates closely with the degree of octahedral substitution.

When the isomorphous replacements occur mainly in the tetrahedral silica layer, as is the case with micas, the cation exchange capacity is much lower, in marked contrast to that found for the montmorillonites (see Figure 1.6). Ions capable of four-fold co-ordination, principally aluminium, may replace silicon in the tetrahedral layer. The resulting charge deficiencies act over a much shorter distance compared with the montmorillonites. Thus the counter-balancing ions are held firmly in the inter-lamellar spaces and serve to bind the layers together by means of strong ionic bonding, and there is no inter-lamellar water of hydration.

In the mica structures the lattice is not capable of expanding, which is clearly evident from its familiar use as a thermal insulation material, and the cations are not exchangeable. The observed exchange capacity parallels that of the kaolin minerals, being almost exclusively dependent on surface area and therefore particle size. Illites are transitional between kaolinites or micas and the montmorillonites and their exchange capacities are impossible to systemize. The characteristic ion exchange properties of the layer lattice aluminosilicates have been reviewed extensively and some typical exchange capacity values, in equivalents per kilogram, or $eq\,kg^{-1}$ (see Chapter 4), appear in Table 1.3.

Table 1.3 *Typical cation exchange capacities of some silicate minerals* (from D. Carrol, *Bull. Geol. Soc. Am.*, 1959, **70**, 754)

Structural type	Mineral	(Approx.) Capacity ($eq\,kg^{-1}$)
Single layer	Kaolinite	0.03–0.15
	Halloysite ($2H_2O$)	0.05–0.10
	Livesite	0.40
Double layer (non-expanding lattice)	Muscovite (mica)	0.10
	Illite (hydrous mica)	0.10–0.40
	Glauconite	0.11–0.20
	Pyrophyllite	0.04
	Talc	0.01
Double layer (expanding lattice)	Montmorillonite	0.70–1.00
	Vermiculite	1.00–1.50
	Nontronite	0.57–0.64
	Saponite	0.69–0.81
Three-dimensional (dense lattice)	Felspar (Orthoclase)	0.02
	Quartz	0.05
(open lattice)	Zeolites	3.0–6.0

Framework Structures

A hitherto unmentioned class of aluminosilicates possessing well defined ion exchange properties comprises those whose structures are a continuous three-dimensional framework lattice. These minerals are important both to the application and to the theoretical understanding of ion exchange processes. In framework silicate structures each of the four oxygen atoms of every silica tetrahedron is linked to two silicon atoms. They receive one valency share from each and thus electrical neutrality is preserved throughout the resulting three-dimensional framework. Theoretically, all structures of this type are modifications of silica $(SiO_2)_n$ but substitution for silicon can occur, giving rise to a variety of minerals containing lattice charge-balancing cations.

Felspars. Felspars (or feldspars) are the commonest example where aluminium atoms substitute for silicon, and the resulting charge deficiency is counterbalanced by alkali or alkaline earth cations contained in holes in the lattice, thus:

$$4\,SiO_2 \rightleftharpoons Si_4O_8 \rightleftharpoons \underset{\text{felspar}}{K.AlSi_3O_8} \qquad (1.3)$$

The potassium or other counterbalancing ions are held quite rigidly within the cavities of the lattice and cannot be exchanged except by disruption of the lattice. The framework structure of a typical felspar is based on condensed rings of four tetrahedra forming a chain-type structure. A moderately open lattice results when this prime unit is linked to similar formations on all four sides.

Zeolites. Zeolites, discovered in 1756, form a most important class of silicate mineral which through their great natural abundance and ready artificial synthesis have assumed wide industrial application. The zeolites are based on a three-dimensional structure with all tetrahedral silica units sharing all their oxygen atoms with other tetrahedra. Isomorphous substitution of silicon by aluminium confers a net positive charge deficiency within the lattice which is balanced by interstitial cations, giving rise to a general stoichiometry given by the empirical formula $M_{2/n}O.Al_2O_3.xSiO_2.yH_2O$, where M is a cation of valency n (commonly $n = 1$ or 2).

The three-dimensional lattice is more open than that of the felspars and in the hydrated form the cations are not firmly held but are free to migrate within the lattice and can be readily exchanged. For example, in the latter part of the nineteenth century, E. Lemberg demonstrated that the zeolite mineral analcite could be converted stoichiometrically and reversibly to leucite simply by leaching with an aqueous solution of potassium chloride:

$$\underset{\text{analcite}}{Na[AlSi_2O_6].H_2O} \underset{NaCl(aq)}{\overset{KCl(aq)}{\rightleftarrows}} \underset{\text{leucite}}{K[AlSi_2O_6]} + H_2O \qquad (1.4)$$

Unlike the layer lattice silicates the rigid open three-dimensional structure of the zeolites prevents any lattice expansion upon water loss under heating, but water loss and regain by the zeolites is quite reversible. The unique lattice structure of the zeolites gives rise to various lamellar and fibrous structures which are characterized by geometrically well-defined channels and cavities, the dimensions of which largely govern the mobility and siting of interstitial ions and their availability for exchange.

The detailed structures of zeolites are varied and complex, but the common pseudo-cubic basket-like frameworks of linked tetrahedra are depicted in Figure 1.7. These typify the unit cage structures found for the mineral faujasite and the synthetic zeolites whose structures are generically described as types A, X, and Y. The constitution, channel diameters, and ion exchange capacities for various natural and synthetic zeolites are shown in Table 1.4.

Natural deposits of zeolites are widespread and widely exploited, but large scale synthesis by controlled hydrothermal crystallization from solutions or precipitation from gels has proved most significant and enables the Si:Al lattice ratio to be modified, giving rise to variations in sorptive activity and ion exchange properties. The precise geometry of the zeolite structures enable them to differentially sorb neutral molecules according to their size or structure, a property which gave rise to zeolites being termed 'Molecular Sieves'. This property is the basis of many industrial uses such as gas and liquid phase operations to effect separation of hydrocarbons, drying, and as catalytic substrates for petrochemical cracking and reforming reactions.

Whilst layer lattice clay minerals are finely divided microcrystalline minerals, which together with their variable composition limits their choice as ion exchangers in industrial processes, the naturally occurring zeolites are largely macrocrystalline. As industrial ion exchangers they are preferable and furthermore they are readily synthesized and in this form may possess even better ion exchange capacities than their naturally occurring counterparts. Artificial zeolites, or

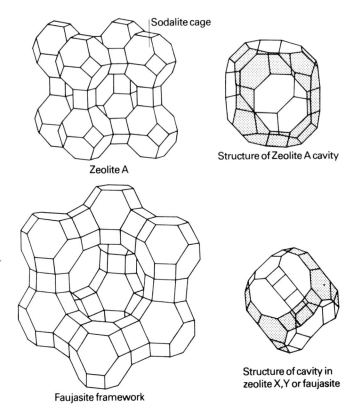

Figure 1.7 *Diagrammatic structure of faujasite and related synthetic zeolites* (Reproduced by permission from J. Dwyer and A. Dyer, *Chem. Ind. (London)*, 1984, 237)

'Permutits' as they were called, were used by Gans for developing the water softening process at the beginning of this century. The word 'Permutit' is derived from the Greek meaning 'to exchange'.

In a way the zeolites are the inorganic counterparts of modern macroporous organic cation exchange resins (see Chapter 3). However, although resins have superseded zeolites in the dominating area of industrial ion exchange associated largely with water treatment, the role of zeolites remains valid for special applications, and in environments which are hostile to resins, for example, high temperatures or ionizing nuclear radiation.

The disadvantages of zeolites over resins for conventional ion exchange applications arise largely from their irregular physical form, friability, slower kinetics, and most importantly their chemical

Table 1.4 *Constitution and ion exchange capacities of some natural and synthetic zeolites* (from H. S. Sherry, in 'Ion Exchange (A Series of Advances)', ed. J. A. Marinsky, Marcel Dekker, New York, 1969, Vol. 2)

Species	Idealized formula	Channel diameter (nm)	Exchange capacity (eq kg^{-1})
Analcite	$Na_{16}[(AlO_2)_{16}(SiO_2)_{32}].16H_2O$	0.69	4.95
Chabazite	$Ca_2[(AlO_2)_4(SiO_2)_8].13H_2O$	0.37–0.42	4.95
Phillipsite	$(K,Na)_{10}[(AlO_2)_{10}(SiO_2)_{22}].20H_2O$	0.42–0.44	4.67
Clinoptilolite	$(Ca,Na_2,K_2)_3[(AlO_2)_6(SiO_2)_{30}].24H_2O$	0.24–0.61	2.64
Mordenite	$Na_8[(AlO_2)_8(SiO_2)_{40}].24H_2O$	0.29–0.70	2.62
Faujasite	$(Na_2,Ca)_{32}[(AlO_2)_{64}(SiO_2)_{128}].256H_2O$	0.74	5.02
Linde A	$Na_{96}[(AlO_2)_{96}(SiO_2)_{96}].216H_2O$	0.41	4.95
Linde X	$Na_{85}[(AlO_2)_{85}(SiO_2)_{107}].256H_2O$	0.25–0.74	6.34
Linde Y	$Na_{52}[(AlO_2)_{52}(SiO_2)_{140}].256H_2O$	0.25–0.74	4.10

instability in solutions of high and low pH. The following chapter deals with the emergence of resinous organic ion exchange materials which, had they not been invented, would have had an enormous detrimental impact upon the technological advancement within the process industries and therefore upon the living standards of virtually everybody.

FURTHER READING

R. W. Grimshaw, 'The Chemistry and Physics of Clays', E. Benn, London, 1971.

R. E. Grim, 'Clay Mineralogy', McGraw–Hill, London, 1953.

R. M. Barrer, 'Zeolites and Clay Minerals', Academic Press, London, 1978.

C. B. Amphlett, 'Inorganic Ion Exchangers', Elsevier, Amsterdam, 1964.

A. Weiss and E. Sextl, 'Clay Minerals as Ion Exchangers', in 'Ion Exchangers', ed. K. Dorfner, Walter de Gruyter, Berlin and New York, 1991, Ch. 1.6, p. 492.

M. Baacke and A. Kiss, 'Zeolites', in 'Ion Exchangers', ed. K. Dorfner, Walter de Gruyter, Berlin and New York, 1991, Ch. 1.5, p. 473.

J. Dwyer and A. Dyer, 'Zeolites – An Introduction', *Chem. Ind. (London)*, 1984, 237.

R. Coffey and T. Gudowicz, 'Detergents – Trends In Siliceous Builders', *Chem. Ind. (London)*, 1990, 169.

H. S. Sherry, 'The Ion-Exchange Properties of Zeolites', in 'Ion Exchange (A Series of Advances)', ed. J. A. Marinsky, Marcel Dekker, New York, 1969, Vol. 2, p. 89.

Chapter 2

The Development of Organic Ion Exchange Resins

The previous chapter dealt with the discovery and properties of aluminosilicate ion exchangers whereby it is immediately apparent that *three fundamental requirements* have to be met to confer ion exchange properties upon a material:

1. An inert host structure which allows diffusion of hydrated ions, *i.e.* a *hydrophilic matrix*.
2. The host structure must carry a fixed ionic charge, termed the *fixed ion*.
3. Electrical neutrality of the structure must be established by the presence of a mobile ion of opposite charge to that of the fixed ion, called the *counter-ion*.

With the above listed essential requirements in mind the ideal characteristics of an ion exchange material listed at the outset of Chapter 1 are reviewed and restated as follows:

1. A *hydrophilic structure* of regular and reproducible form.
2. A controlled and effective *exchange capacity*.
3. A *reversible* and *rapid rate* of exchange.
4. *Chemical stability* towards electrolyte solutions.
5. *Physical stability* in terms of mechanical strength and resistance to attrition.
6. *Thermal stability*.
7. *Consistent particle size* and effective surface area compatible with the hydraulic design requirements for industrial scale plant.
8. *An option on the type of exchanger* so as to be able to select either *cation or anion* exchange.

With regard to the requirements 4, 5, 7, and 8 listed above the aluminosilicate materials were less than ideal which prompted investigations to seek alternatives.

EARLY ORGANIC ION EXCHANGE MATERIALS

Natural Products

Once the nature of ion exchange had been established by experiments with aluminosilicate exchangers, the potential of other materials, particularly certain organic substances, was realized. Many substances of an organic nature were examined but the first real success was with sulfonated coals around 1900. Certain types of soft coal, when treated with hot fuming sulfuric acid, reacted partially to give sulfonic acid groups (—SO_3H) on the hydrocarbon matrix. The sulfonic acid groups conferred a measure of hydrophilic nature so that when placed in aqueous solution the material ionized according to the equation:

$$R(SO_3H)_n \overset{H_2O}{\rightleftharpoons} [R(SO_3)_n]^{n-} + n\,H^+ \qquad (2.1)$$

where R represents the hydrocarbon matrix, and n is the number of fixed ionizable sulfonic acid groups carried on the matrix. Thus the hydrogen counter-ions are able to exchange for cations in the external solution.* These materials were termed 'carbonaceous' exchangers or 'carbonaceous zeolites' from which the trade name 'Zeo-Karb' was coined by The Permutit Company in the 1930s. This class of exchanger lacked uniformity, physical and chemical stability, and possessed about one third the total ion exchange capacity of modern resins. Yet with all these shortcomings they were in use as cation exchangers in industrial water treatment applications until as late as the mid-1970s. Now though, they can be regarded as obsolete.

Condensation Polymers

The first completely synthetic ion exchange resins were prepared by B. A. Adams and E. L. Holmes in 1935. The basis of their synthesis was the condensation polymerization of methanal (formaldehyde) with phenol or polysubstituted benzene compounds to give, after

---

* Unless otherwise stated the hydrated hydroxonium ion H_3O^+ is implied but written H^+ for convenience.

crushing and grading, a brittle granular resin similar to 'Bakelite' in appearance.

A generic synthesis route is shown schematically in Scheme 2.1 but the underlying principles are best understood by considering the reaction in separate stages. For example, with phenol and methanal reaction occurs under heat to give a phenol–methanal *polymer* chain with the elimination of water (hence the term 'condensation'). At the same time condensation occurs between the propagating polymer chains and methanal to give a *crosslinked copolymer*. At this stage the

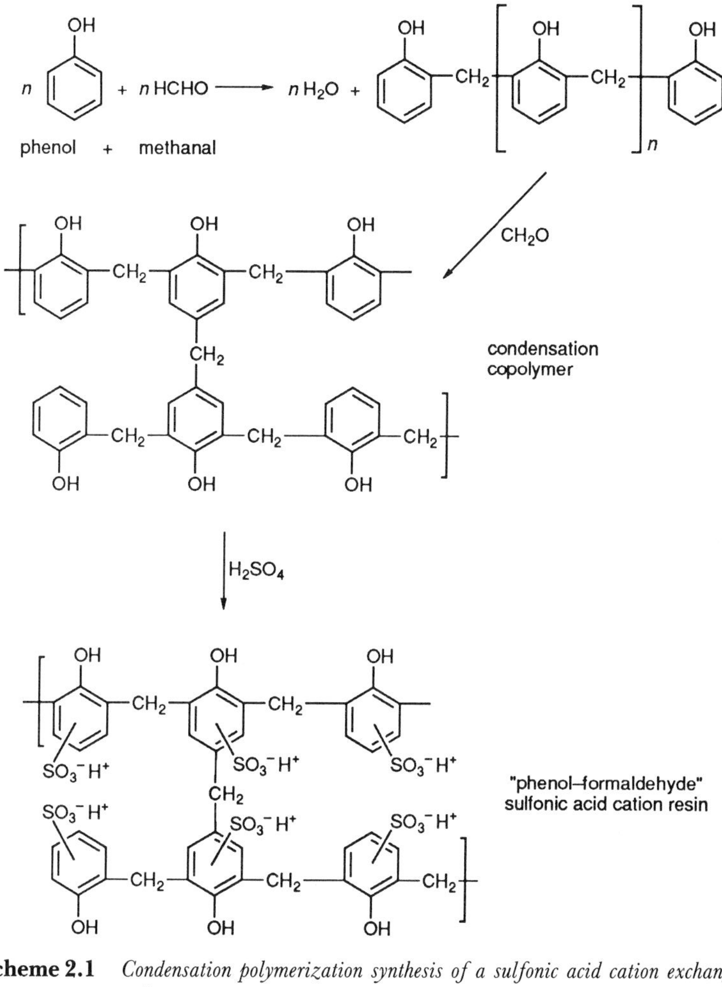

Scheme 2.1 *Condensation polymerization synthesis of a sulfonic acid cation exchange resin*

copolymer has no ion exchange property except that arising from the phenolic hydroxyl group (—OH) which being very weakly acidic would only ionize at very high pH. The next stage is to introduce the ion exchange *functional group*, namely sulfonic acid (—SO$_3$H), by reaction with hot sulfuric acid. Often it is possible to incorporate the functional, or *ionogenic*, group at the copolymerization stage by using an appropriately substituted benzene reactant, for example, 3-hydroxybenzenesulfonic acid (metaphenolsulfonic acid). Polymer chain propagation and crosslinking occur simultaneously but at different rates giving rise to an essentially heterogeneous structure. Cross-linking is essential since otherwise a linear polymer would be produced whereupon the derived ion exchanger would be soluble. For the example cited, the resulting material is described as a 'phenol–formaldehyde sulfonic acid' cation exchanger, RSO$_3$H, which in aqueous solution dissociates thus:

$$RSO_3H \rightarrow RSO_3^- + H^+ \tag{2.2}$$

where R represents the copolymer matrix and the sulfonate group (—SO$_3^-$) the fixed anion. Once the functional group has been defined, the dissociation in an aqueous environment can be more simply represented thus:

$$RH \rightarrow R^- + H^+ \tag{2.3}$$

These highly significant developments related to cation exchange resin synthesis were quickly followed up by Adams and Holmes with an analogous synthesis route for anion exchange resins. This involved condensation polymerization between methanal and various phenylamines, giving directly a copolymer matrix carrying weakly basic secondary amine groups which may be simplistically denoted R$_2$NH. At first sight it is not readily apparent how such a grouping functions as an ion exchanger. Its ion exchange characteristics arise from the property of weak base primary, secondary, and tertiary amines to undergo an addition reaction with aqueous solutions of strong acids to give acid amine salts. For example, with hydrochloric acid:

$$R_2NH + HCl \rightarrow R_2NH_2^+ Cl^- \tag{2.4}$$

Thus once the required configuration of a fixed cation and mobile (exchangeable) counter-anion is achieved exchange may proceed conventionally with another acid electrolyte, for example:

$$2\,R_2NH_2{}^+\,Cl^- + H_2SO_4 \rightleftharpoons (R_2NH_2{}^+)_2\,SO_4{}^{2-} + 2\,HCl \quad (2.5)$$

The basic synthesis steps of copolymer formation with the *in situ* inclusion of the functional group or its subsequent introduction into the matrix is the basis of all commercial ion exchange resin production including modern day products. The realization of synthetic routes for both cation and anion exchange resins immediately widened the field of potential applications and so heralded the beginning of the technological advancement of ion exchange to become the chemical engineering unit process we know today. Even so, the early condensation polymers still lacked ideal physical and chemical stability and have now largely given way to the advanced products which constitute today's modern resins.

MODERN ORGANIC ION EXCHANGE RESINS

Addition Polymerization

The basis of modern organic resin production incorporates the same principles described for their predecessors but depends upon an entirely different polymerization mechanism first applied by D'Alelio in 1944, called *addition* or *vinyl polymerization*. The mechanism is one of free radical induced polymerization between reactants (*monomers*) carrying ethenyl (or vinyl) double bonds ($-CH=CH_2$). One of the reactants must contain at least two ethenyl double bonds to effect crosslinking. Again, an understanding of the resin synthesis is afforded by showing separately in Scheme 2.2 what are in fact simultaneously occurring complex polymerization reactions between all reactant permutations.

Styrenic Cation Exchange Resins

The miscible monomers, ethenylbenzene (styrene) and diethenylbenzene (divinylbenzene, DVB), undergo a free radical induced copolymerization reaction initiated by a benzoyl peroxide catalyst. The exothermic reaction is carried out in an aqueous suspension whereby the mixed monomers are immiscibly dispersed as spherical droplets throughout the reacting medium resulting in discreet beads of copolymer being formed. Correct reaction conditions and the use of suspension stabilizers enable the particle size distribution of the

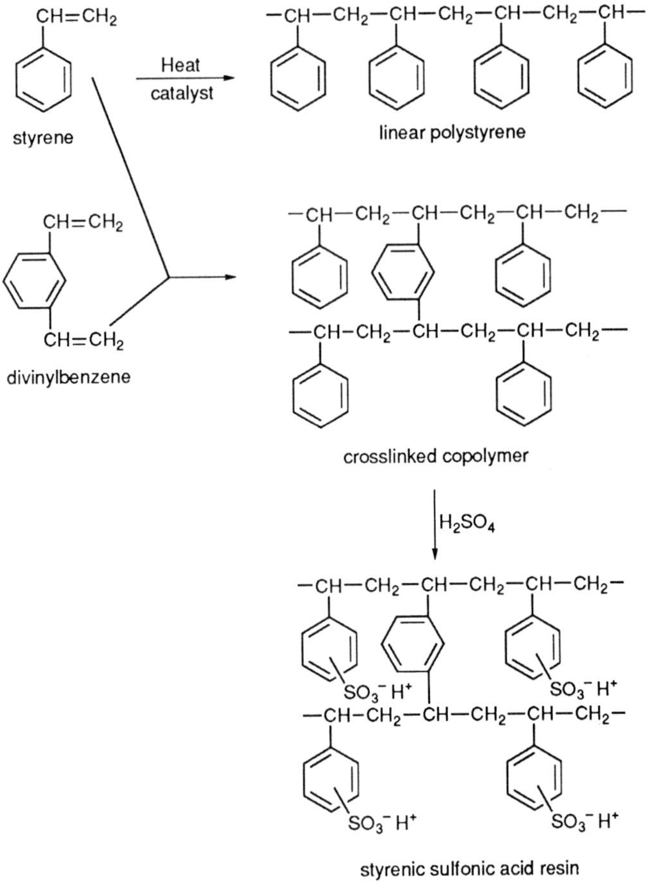

Scheme 2.2 *Addition polymerization synthesis of a styrene sulfonic acid cation exchange resin*

copolymer to be closely controlled. The extent to which the copolymer is crosslinked depends upon the proportion of crosslinking agent (divinylbenzene) employed in the synthesis and has a pronounced impact upon both the mechanical and chemical behaviour of the derived ion exchange resin.

Activation of the copolymer is carried out by sulfonation of the matrix with hot sulfuric acid thereby introducing the sulfonic acid functional group giving a *strongly acidic* cation exchange resin. The reaction in Scheme 2.2 shows sulfonic acid substitution occurring within the 'styrene' nucleus only, but whether or not all aromatic nuclei become sulfonated is a subject of some debate. Subsequent treatment of the sulfonic acid resin (RSO_3H) with brine or sodium hydroxide solution gives, via ion exchange, the sodium sulfonate salt form (RSO_3Na). Finally rinsing and grading produces the now so familiar bead form product characteristic of addition polymerized resins (Figure 2.1).

Acrylic Cation Exchange Resins

Ethenylbenzene is not the only ethenyl bonded monomer capable of undergoing copolymerization with diethenylbenzene (divinylbenzene), but commercially, the propenoic (acrylic) monomers are the alternatives which have been most widely exploited, since about 1950. For example, the 'methacrylic–divinylbenzene' *weakly acidic* cation exchange resin [—$RC(CH_3)COOH$] is made by copolymerizing diethenylbenzene and methylpropenoic acid (methacrylic acid) as shown in Scheme 2.3. Various alkyl substituted propenoic acid monomers may be employed in the manufacture of weakly acidic cation exchange resins, as are propenonitriles (acrylonitriles) and alkyl propenoates (acrylic esters). In the case of the two latter cited

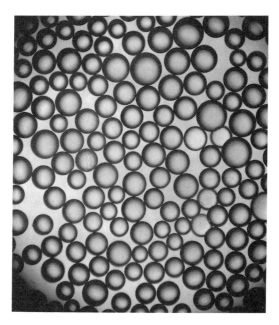

Figure 2.1 *The typical bead appearance of modern addition polymerized ion exchange resins*

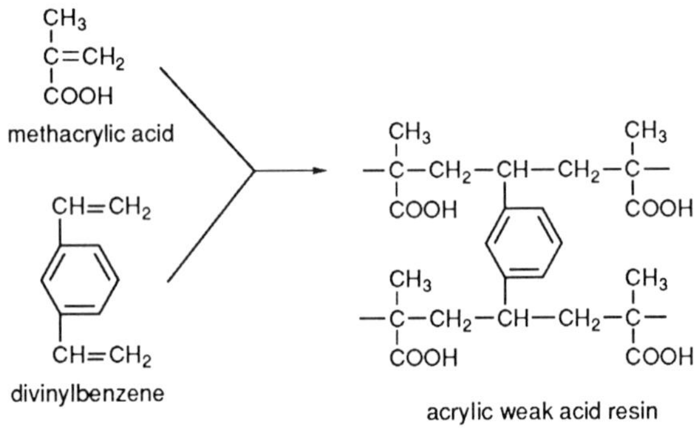

Scheme 2.3 *Addition polymerization synthesis of an acrylic carboxylic cation exchange resin*

monomers the derived copolymer is further subjected to an acid hydrolysis stage to give the carboxylic acid functional group, as illustrated by the equations:

$$\begin{aligned} R\text{—}CN + 2\,H_2O &\xrightarrow{\text{acid}} R\text{—}COOH + NH_3 \\ R\text{—}COOR' + H_2O &\xrightarrow{\text{hydrolysis}} R\text{—}COOH + R'OH \end{aligned} \quad (2.6)$$

(where R' is an alkyl group)

A *bifunctional* cation exchange resin carrying strongly acidic (sulfonic) and weakly acidic (carboxylic) groups was introduced in the mid-1960s but was to never merit a sustained commercial viability, and has since been discontinued. Undoubtedly, the 'acrylic' and 'styrene' copolymers with 'divinylbenzene' form the basis of most commercially manufactured cation exchange resins available today.

Styrenic Anion Exchange Resins

The suspension polymerized ethenylbenzene–diethenylbenzene copolymer is also the host matrix for most anion exchange resins. The preformed copolymer is subject to two further synthesis steps, first developed by McBurney in 1947, as described below and by Scheme 2.4.

1. *Chloromethylation.* A *Friedel–Crafts* reaction between the copolymer and chloromethoxymethane with aluminium chloride as the catalyst introduces chloromethyl groups (—CH_2Cl) into the ethenylbenzene nuclei. What appears to be a simple step is in fact a critical stage in the synthesis, demanding strict techniques to firstly minimize undesir-

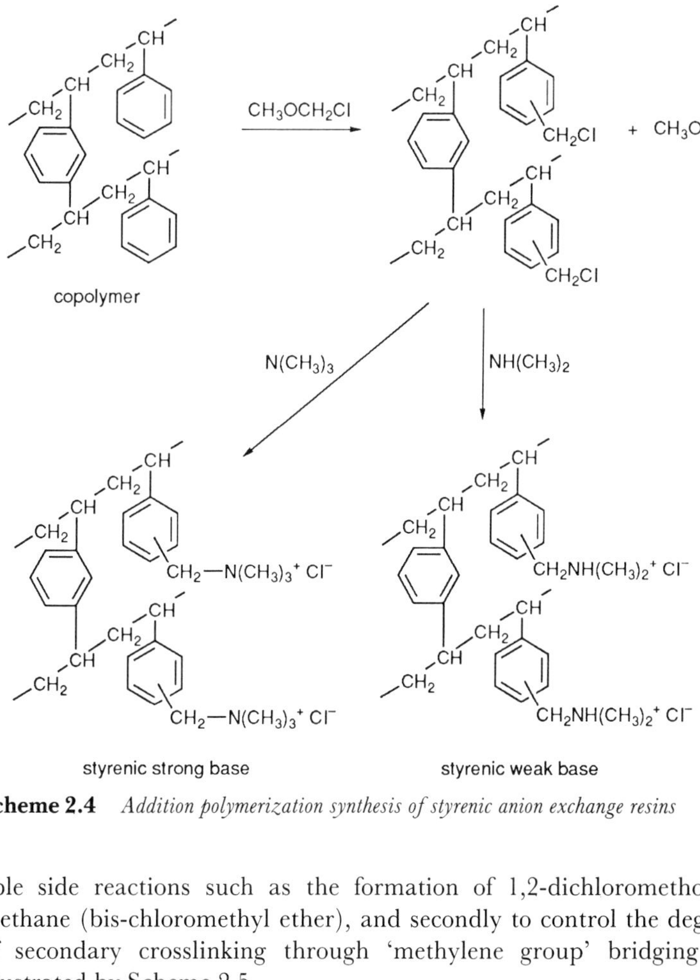

Scheme 2.4 *Addition polymerization synthesis of styrenic anion exchange resins*

able side reactions such as the formation of 1,2-dichloromethoxy-methane (bis-chloromethyl ether), and secondly to control the degree of secondary crosslinking through 'methylene group' bridging as illustrated by Scheme 2.5.

2. *Amination.* The final stage after purification of the chloromethylated copolymer is the substitution of the functional group by reaction with various alkyl substituted aliphatic amines as shown in Scheme 2.4. Trimethylamine, $(CH_3)_3N$, gives the quaternary benzyltrimethylammonium chloride functional group, $RCH_2N(CH_3)_3^+ Cl^-$, which is characteristic of most *Type I strongly basic* anion exchange resins. The equivalent reaction using dimethylethanolamine, $(CH_3)_2(C_2H_4OH)N$, gives the *Type II* class of strong base anion exchange resins, $RCH_2N(CH_3)_2(C_2H_4OH)^+ Cl^-$. If instead of using tertiary trimethylamine, methylamine or dimethylamine is employed, the resulting

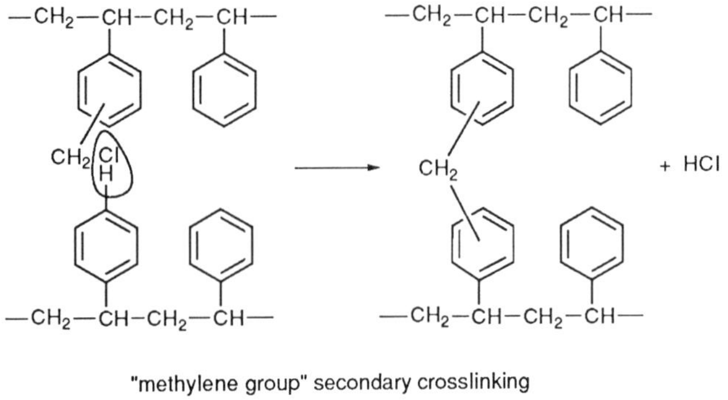

"methylene group" secondary crosslinking

Scheme 2.5 *Secondary crosslinking through 'methylene group' bridging*

resins are *weakly basic* with secondary, $RCH_2NH(CH_3)$, or tertiary, $RCH_2N(CH_3)_2$, functionality.

The range of amine functional group configurations is quite large because of the many suitable copolymers and amine derivatives available. However, in the main, most commercially available styrene based anion exchange resins are based on weakly basic secondary and tertiary amine functional groups or the strongly basic quaternary ammonium grouping.

Acrylic Anion Exchange Resins

It was predicted, and subsequently confirmed, that an anion exchange resin based upon an acrylic matrix should demonstrate beneficial exchange equilibria and kinetics towards large organic ions compared with the styrenic structures. A simplistic explanation for this lies with the greater hydrophilic nature of the aliphatic skeletal structure of the acrylic matrix, which in turn means a weaker van der Waals type attraction between the resin matrix and the hydrocarbon structure of an organic counter-ion.

The practical implications of this property are highlighted in later chapters, and it suffices to state at this stage that from the mid-1960s onwards the full compliment of acrylic anion exchangers were developed: weak base, strong base, and bifunctional. Commonly methyl propenoate (methyl acrylate) is chosen as the monomer for copolymerization with diethenylbenzene to give the host matrix as shown in Scheme 2.6.

1. *Weak Base Functionality*. Amination with dimethylaminopropylamine (DMAPA) introduces a tertiary amine functional group.

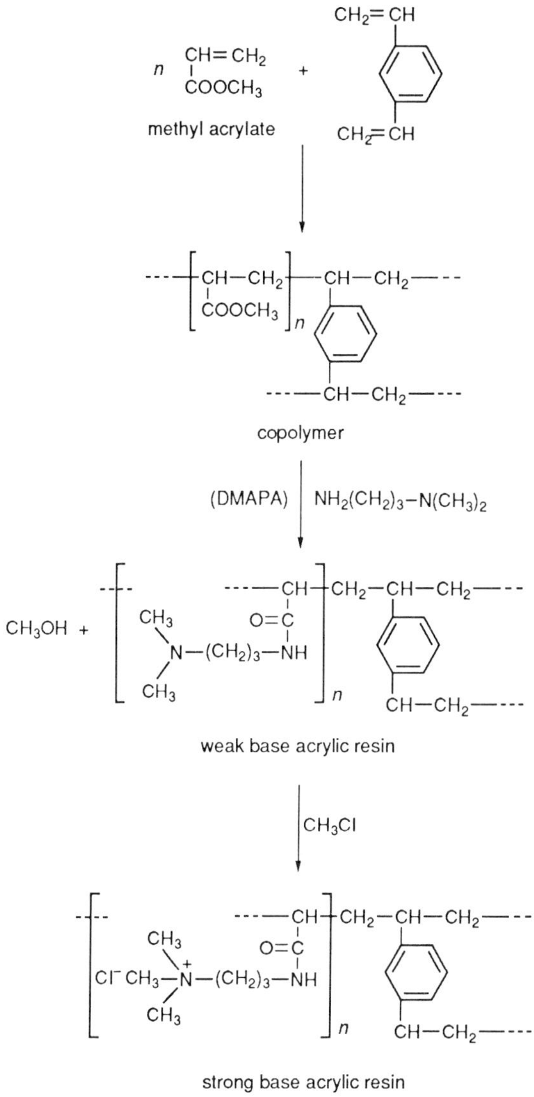

The Development of Organic Ion Exchange Resins

Scheme 2.6 *Addition polymerization synthesis of acrylic anion exchange resins*

2. *Strong base functionality.* As shown in Scheme 2.6 a subsequent 'quaternization' step employing chloromethane (methyl chloride) converts the weak base product to the strongly basic quaternary ammonium resin.

3. *Bifunctionality*. Although bifunctional properties may be obtained by physically mixing weak and strong functional resins, a truly bifunctional exchanger is one which contains both types of group within the same bead. Anion exchange resins possessing true bifunctionality within an acrylic matrix are well established, but the analogous cation exchange resin remains unavailable commercially.

SPECIAL ION EXCHANGE MATERIALS

The vast majority of applications and theoretical treatments of ion exchange are concerned with the established copolymer materials previously described. However, an introduction to resin exchangers would be incomplete without a brief reference to some other more specialized products.

Specific Ion Exchangers

If an ion forms strong complexes with, or is precipitated by, a certain class of chemical reagent, ion exchange resins incorporating such a class of compound as its functional group usually exhibit a high affinity for such ions. Such an exchanger which strongly takes up one (or at least not more than a few) counter-ion species relative to all others has great potential usefulness in analytical chemistry, and as a possible basis of commercial recovery or purification processes. The first specific cation exchanger was patented by A. Skogseid in 1947 containing functional groups similar to 2,2',4,4',6,6'-hexanitrodiphenylamine and showed a specific affinity for potassium ions.

The chemistry of the common heavy metals is characterized by their readily forming co-ordination complexes or chelates with electron pair donating ligands. Therefore it is not surprising that a styrenic exchanger containing iminodiacetate functional groups should show particularly strong affinity towards many polyvalent and transition metal cations. The type of chelate structure to be expected with such a resin is illustrated by Figure 2.2 along with that for the analogous ethylenediaminetetraacetic acid (EDTA) complex for comparison. Numerous specific ion exchangers have been reported, but those more commonly encountered are listed in Table 2.1.

It is important to realize that specificity does not necessarily mean an affinity by the exchanger for one ion only, and in many ways the term 'specific' can be misleading. Selective or chelating ion exchange is a better description since commonly the relative affinities of the resin for several ions, not just one ion, are enhanced compared with

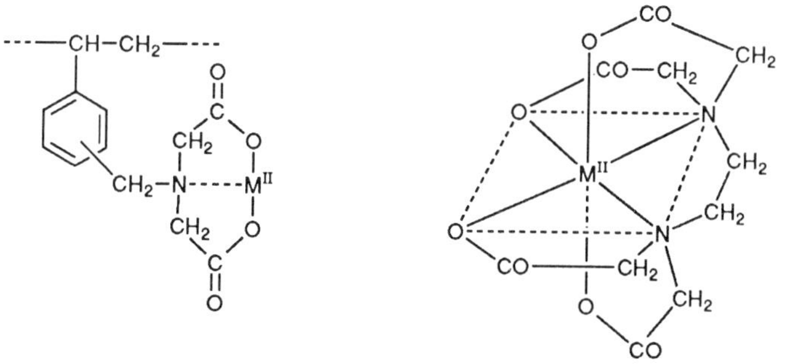

Iminodiacetate chelate

EDTA chelate (a suggested structure)

Figure 2.2 *An example of a specific ion exchange resin and its relation to an analytical reagent*
(Adapted from R. W. Grimshaw and C. E. Harland, 'Ion-Exchange: Introduction to Theory and Practice', The Chemical Society, London, 1975)

Table 2.1 *Matrix, functional group, and ion affinities of some common specific ion exchangers*

Matrix	Ionogenic Group	Specificity
styrene–DVB	iminodiacetate $—CH_2—N(CH_2COO^-)_2$	Fe, Ni, Co, Cu; Ca, Mg
styrene–DVB	aminophosphonate $—CH_2—NH(CH_2PO_3)^{2-}$	Pb, Cu, Zn, UO_2^{2+}; Ca, Mg
styrene–DVB	thiol; thiocarbamide $—SH; —CH_2—SC(NH)NH_2$	Pt, Pd, Au; Hg
styrene–DVB	N-methylglucamine $—CH_2N(CH_3)[(CHOH)_4CH_2OH]$	B, (as boric acid)
styrene–DVB	benzyltriethylammonium $—CH_2N(CH_2CH_3)_3^+$	NO_3^-
phenol–formaldehyde	phenol; phenol-methylenesulfonate $—C_6H_3(OH), —C_6H_2(OH)CH_2SO_3^-$	Cs

conventional exchangers. For example the aminophosphonic chelating resins are highly selective towards divalent alkaline earth cations over monovalent ions; and certain types of quaternary benzyltrialkylammonium strong base anion exchangers are more selective for the nitrate ion over the sulfate ion which is a reversal of the normal sequence. Both these examples form the basis of recent process applications which are discussed in Chapter 8.

It is equally important to appreciate that a high selectivity for a particular ion does not necessarily mean that the resin concerned is bound to have immediate commercial application. The reason for this is that the property of high affinity is always associated with a reduction in the degree of reversibility in cyclic operations thereby rendering regeneration difficult.

Tailored copolymer resins are not the only exchangers to exhibit specific affinities towards selected ions. Many types of inorganic materials such as clays, zeolites, amphoteric oxides, heteropolyacid salts, and phosphates exhibit useful specificity towards selected monovalent and polyvalent ions. In the laboratory such media are often the basis of chromatographic separations, whilst industrially many such materials offer benefits in radioactive waste effluent treatment for removing nucleides such as caesium (^{137}Cs) and strontium (^{90}Sr).

Membrane Materials

About 30 years ago attention was focused on the development of ion exchange polymers in the form of membranes. Their main use is as permselective membranes allowing the electrochemical transport of ions of one charge type only, depending upon whether the membrane comprises a cation or anion exchange resin. Besides being the objects of pure scientific study, these materials have a great potential application in the general field of electrochemistry and specifically saline water treatment. Electrodialysis plants using ion exchange membranes for desalting brackish waters are now operating commercially. Heterogeneous membranes comprising an inert binding material impregnated with a suitable finely-divided ion exchange resin have been prepared, but these have been superseded by improved homogeneous membranes which may be regarded as a 'sheet like' analogy to conventional modern ion exchange resins.

Liquid Exchangers

An ion exchange reaction is usually considered as taking place between a solid and a liquid phase. A degree of flexibility in this definition is required in order to accommodate the various organic liquid ion exchangers which have found important application over the last 40 years, especially in the field of extraction metallurgy. The mode of operation of liquid ion exchangers is analogous to that of the resinous materials, except that, in the former case, exchange occurs at the phase boundaries formed between two immiscible liquids such as kerosene, containing the exchanger, and an aqueous electrolyte. On acquiring the desired degree of extraction into the organic phase, the exchange reaction may be reversed by a repeated contact or scrubbing of the organic phase with a concentrated solution of a suitable electrolyte.

The most successful anion exchangers of this type are high molecular weight amine derivatives, whilst as cation exchangers, organo-phosphoric and carboxylic acids have proved particularly successful. Liquid exchangers have become established as liquid–liquid extraction reagents for separating nuclear fission products and recovering metals such as copper and uranium from leach liquors. Conventional ion exchangers are also used for such operations and are discussed in Chapter 9.

Amphoteric Exchangers

Amphoteric ion exchange resins contain both acidic and basic functional groups. Various exchangers of this type have been prepared, but applications have been found for only a few, of which the most important are the 'snake-cage polyelectrolytes'. These materials are conventional cation or anion exchangers containing polymerized counter-ions of opposite charge to the fixed ion which are permanently entangled with the crosslinked matrix of the host exchanger. Resins of this type exhibit simultaneous cation and anion exchange behaviour. However, as discussed in Chapter 5, the affinity of an exchanger towards ions differs according to the particular ion concerned. Therefore resins of this type have an application as reversible selective sorbants for electrolytes sharing a common ion, and serve to facilitate the separation of electrolyte mixtures by a principle known as 'ion-retardation'.

Oxidation–Reduction Resins

Expertise in resin and polymer synthesis has made available materials which can function as solid, but insoluble, oxidizing or reducing agents; they are sometimes called redox resins or 'electron exchangers'. Their reactivity in the latter sense is due to the polymer matrix carrying functional groups which may be reversibly oxidized and reduced. Studies of these polymers have involved two main kinds of approach. In the first instance polymers incorporating the quinone–hydroquinone redox couple and their derivatives have been investigated. The alternative approach has been to use conventional strong acid and strong base resins as substrates for redox-active cations and anions respectively.

Although these materials are still too costly to be used in commercial processing they do possess potential in this respect and may be employed as reagents in redox titrations. An interesting exception is afforded by the mineral glauconite which when impregnated with manganese(IV) oxide acts as a redox couple, thus:

$$Mn^{IV} + 2\,e^- \rightleftharpoons Mn^{II} \qquad (2.7)$$

The oxidation of soluble iron(II) and manganese(II) hydrogencarbonates to insoluble iron(III) and manganese(IV) oxides using a manganese 'zeolite' filter is the basis of a much used process for iron and manganese removal in water treatment. A simplified representation of the somewhat complex redox reaction are given by the following equations:

$$\left. \begin{array}{l} Fe^{II} - e^- \rightarrow Fe^{III} \\ Mn^{II} - 2\,e^- \rightarrow Mn^{IV} \end{array} \right\} \text{Oxidation} \qquad \begin{array}{l}(2.8)\\(2.9)\end{array}$$

Overall:

$$Mn^{II} + 2\,Fe^{II} + \underset{\text{'glauconite'}}{2\,Mn^{IV}} \rightarrow 2\,Fe^{III} + \underset{\text{'glauconite'}}{2\,Mn^{II}} + Mn^{IV} \qquad (2.10)$$

or

$$2\,Fe^{II}(HCO_3)_2 + \underset{\text{'glauconite'}}{Mn^{IV}O_2} \rightarrow Fe^{III}{}_2O_3(s) + \underset{\text{'glauconite'}}{Mn^{II}O} + 4\,CO_2 + 2\,H_2O \qquad (2.11)$$

and

$$Mn^{II}(HCO_3)_2 + \underset{\text{'glauconite'}}{Mn^{IV}O_2} \rightarrow Mn^{IV}O_2(s) + \underset{\text{'glauconite'}}{Mn^{II}O} + 2\,CO_2 + H_2O \qquad (2.12)$$

The manganese(IV) state once reduced is restored by regenerating the media with potassium permanganate solution. Thus one has an example of a much used redox reaction using, compared with resin, a relatively inexpensive inorganic 'electron exchanger'.

Non-conventional Resins

This classification of products is intended to introduce more recently available ion exchange resin products that are not encountered in the normal macroscopic granular or bead form. For example finely *powdered* cation and anion exchange resins (approx. 50 μm diameter) find application in pharmaceutical formulations as controlled release drug carriers and as swelling agents to assist tablet disintegration and dissolution. Also mixed powdered cation and anion exchange resins are used as precoat filter media to simultaneously effect filtration and ion exchange.

Exchangers of extremely small particle size possess a high specific surface area which greatly accelerates the speed of reaction. This same kinetic enhancement applies to ion exchange papers made by impregnation with ion exchange resin microspheres (0.01 μm–2 μm diameter), and to the *pellicular* resins which are characterized by having a thin ion exchange copolymer film or latex, bonded to the surface of an otherwise inert micro-bead. This latter class of resins, because of their high specific surface area, low capacity, and short diffusion paths have revolutionized ion exchange chromatography techniques whereby complete resolution of only micro amounts of ion mixtures may be achieved in a matter of minutes.

Interest is currently being shown in woven fibrous ion exchange materials which as well as having potential use in analytical separations may also offer process benefits in such areas as effluent treatment and hydrometallurgy.

FURTHER READING

R. M. Wheaton and M. J. Hatch, 'Synthesis of Ion-Exchange Resins', in 'Ion Exchange (A Series of Advances)', ed. J. A. Marinsky, Marcel Dekker, New York, 1969, Vol. 2, Ch. 6, p. 191.

K. Dorfner, 'Synthetic Ion Exchange Resins', in 'Ion Exchangers', ed. K. Dorfner, Walter de Gruyter, Berlin and New York, 1991, Ch. 1.2, p. 189.

R. Kunin, 'Ion Exchange Resins', Krieger, Melbourne, Florida, 1972.

R. Kunin, 'The Synthesis of Ion Exchange Resins', in 'Ion Exchange

Resins', Wiley, New York and London, 2nd Edition, 1958, Ch. 5, p. 73.

A. Warshawsky, 'Chelating Ion Exchangers', in 'Ion Exchange and Sorption Processes In Hydrometallurgy', Critical Reports on Applied Chemistry, Vol. 19, ed. M. Streat and D. Naden, Wiley, London, 1987, p. 166.

Chapter 3

Structure of Ion Exchange Resins

Polymer science continues to play an important role in the development of new and improved ion exchange resins. Furthermore the current 'state of the art' designs and practices in ion exchange technology owe much to the investment made in resin synthesis research and development by leading resin manufacturing companies. The previous chapter, whilst giving an insight into resin synthesis, does not imply that a detailed knowledge of polymer synthesis is essential to understanding ion exchange in the more practical sense. Much more important is an appreciation of firstly resin *structure* since this influences greatly the equilibrium, kinetic, and physical characteristics of a resin. Secondly, and of equal practical importance, is an understanding of the manufacturer's *specification* of a given resin in order to decide selection criteria for any intended application.

CONVENTIONAL RESIN STRUCTURE

Styrenic Gel Resins

The synthesis of the copolymer (matrix) occurs through polymerization between, and in order of decreasing reaction rate:

1. divinylbenzene with divinylbenzene–*rapid*
2. styrene with divinylbenzene–*intermediate*
3. styrene with styrene–*slow*

Because of the differences in polymerization reaction rates the copolymer first formed is greatly crosslinked (*entangled*), but as the reaction proceeds, and the crosslinking agent (divinylbenzene) is consumed, the structure becomes less crosslinked and consequently more open in configuration. Besides purely reaction rate considerations, there is also the fact that in the absence of a solvent for the growing copolymer, except the monomers themselves, the ability of the copolymer to swell diminishes as the reaction proceeds.

Thus the combined reaction rate and steric hindrance effects give rise to a resin which is extremely heterogeneous in structure, varying in crosslinking between very entangled (highly crosslinked) and very open (low crosslinking). Given this heterogeneity with regard to the distribution of crosslinks and therefore functional groups the resin phase is otherwise homogeneous and without discernible porosity. Such a structure is termed *gel-heteroporous* or *gel-microporous*, and of a structure with respect to crosslinking shown schematically in Figure 3.1a. The term 'heteroporous' is somewhat misleading in that there are no actual internal macroscopic structural pores (holes or channels) as evident from Figure 3.2a. Instead the hydrated resin phase,

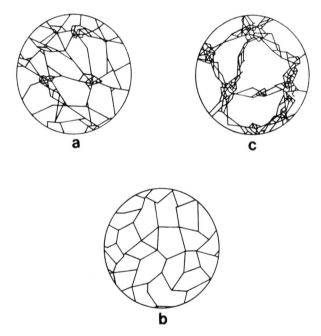

Figure 3.1 *Schematic representation of resin structures*
    a. Gel-microporous
    b. Gel-isoporous
    c. Macroporous

(Reproduced by permission from T. R. E. Kressman, *Effluent & Water Treat. J.*, 1966, **6**, 119)

Structure of Ion Exchange Resins 41

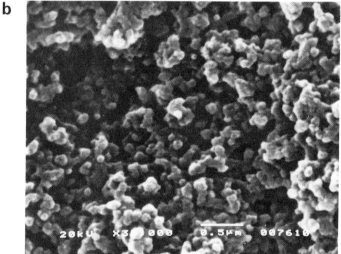

Figure 3.2 *Electron micrographs of resin structure*
    a. Gel resin
    b. Macroporous resin
(Photographs kindly made available by Purolite International Limited)

once ionized, can be likened to a dense electrolyte–gel within which the *dissociated* counter-ions are able to diffuse (Figure 3.3).

The non-uniform distribution of crosslinking and functional groups gives rise to regions showing different structural characteristics within

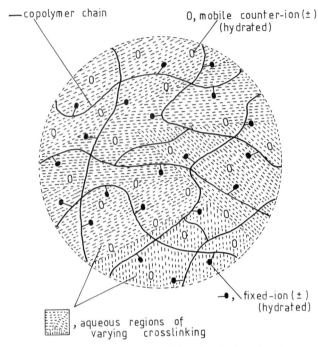

Figure 3.3 *Diagrammatic representation of the resin gel–electrolyte phase*

a given resin bead, and consequently non-symmetrical strains within the structure. Some of the most severe stresses and strains within a resin are sterically induced during activation of the copolymer with the functional group and the subsequent aqueous conditioning and rinses which complete the transition from a hydrophobic to a hydrophilic structure. Also upon hydration the ion exchange resin swells differentially because of the unsymmetrical distribution of crosslinking thereby creating further strain within the resin beads. This is clearly evident when the beads, whilst in the process of swelling, are viewed under a microscope using transmitted polarized light which reveals transient strain patterns (birefringence) as shown in Figure 3.4 and described in Box 3.1.

In order to relieve the strains imposed during activation of a resin it is usual to pre-swell the copolymer with a solvent such as dichloroethene (ethylene dichloride) thereby easing greatly the steric resistance to activation of what would otherwise be a collapsed copolymer structure. By and large resin producers do not publicize, understandably, exact details concerning structure enhancement techniques employed during resin manufacture. Suffice it to say that compared with

Structure of Ion Exchange Resins

BOX 3.1 Demonstration of Structural Strain during Resin Swelling

The birefringence patterns seen under crossed polars as a gel resin swells are easily demonstrated using a polarizing microscope and ordinary transmitted light.

However a simple microscope may also be used given a degree of simple improvization. Obtain some polaroid sheet and secure a piece beneath the stage between the sample slide and light source (or reflecting mirror) as indicated in the Figure.

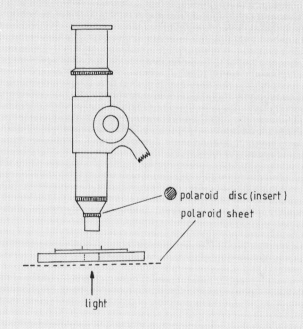

Detach the objective lens housing (or some other convenient lens housing) and cut a circle of polarized sheet to give a snug fit within the housing and covering the lens. After reassembling the lens housing view a slide sample of dried gel cation resin beads ($\geqslant 8\%$ DVB) contained within a shallow plastic or metal washer. Slowly unscrew the lens housing containing the polarizing disc until the field of view is dark (crossed polars). Wet the resin with water (or any other swelling agent) and quickly readjust the focus. Strain within the swelling copolymer structure will show up as bright 'maltese cross' like patterns (shown in Figure 3.4) for as long as it takes for the strain to dissipate – permanent strain would show a permanent pattern.

This experiment suggests a way of studying the swelling and crosslinking characteristics of different resins, for example by measuring the decay times for birefringence patterns.

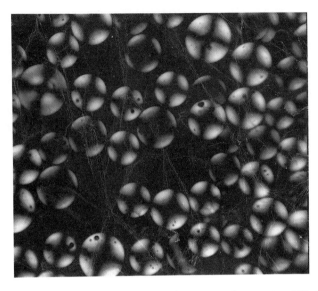

Figure 3.4 *Strain patterns within a swelling gel cation exchange resin – 12% DVB*

the early gel resins many of today's premium products are essentially 'strain free' and possess excellent structural integrity, but some intrinsic heterogeneity remains.

In the 1960s and 1970s the problem of '*organic fouling*' of anion exchange resins (see Chapter 8) greatly impaired the ability of the ion exchange water demineralizing process to give, at acceptable cost, a final treated water quality sufficiently good to meet the increasingly stringent requirements of the process and power industries. Research within academic institutions and engineering companies established definite links between the organic fouling problem and the heterogeneity of gel anion resin structure, thereby fuelling a debate which was to rage for many years culminating in three very significant advances concerning resin structure.

The observed anion resin fouling by large, high molecular weight, aromatic organic ions of natural origin (humic and fulvic acids, Figure 8.8) was established as being due to their becoming entangled within the densely crosslinked and tortuous regions of the anion resin structure. Styrenic gel anion exchange resins tend to be more crosslinked than their divinylbenzene content would suggest due to additional crosslinking through 'methylene group' bridging (see Chapter 2, Scheme 2.5). The latter mechanism, by itself being totally random, would establish a uniform distribution of crosslinks. Therefore a careful manipulation of the crosslinking reaction conditions to yield a total or partial prominence of 'methylene' bridging over solely divinylbenzene crosslinking gives a copolymer which is much more uniformly crosslinked as represented schematically in Figure 3.1b.

Such a structure is termed *gel-isoporous* and does demonstrate a much improved resistance to 'organic fouling' compared with distinctly gel-heteroporous, but otherwise equivalent, resins. Nowadays the term 'isoporous' has largely disappeared by way of the fact that minimizing the degree of heterogeneity of structure is inherent in the production of most modern premium gel anion exchange resins.

SOLVENT MODIFIED RESIN STRUCTURES

Whilst the late 1940s and 1950s realized the foundation of commercial ion exchange resin production with regard to polymer matrix and functionality, the period immediately following heralded major advances with regard to resin structure and physical properties. At the same time as 'isoporous' copolymer synthesis methods were focused on reducing the heterogeneous characteristics of styrenic gel resins, other workers and establishments were adopting an entirely different approach.

As for the 'isoporous' stance, the developments were driven by the need to overcome the 'organic fouling' phenomenon. Accepting that the release of large fouling organic anions from heteroporous gel anion exchange resins was impeded in part through entanglement within the structure, then such release should be facilitated by the resin possessing genuine porosity. In other words, the creation of pores and channels within the resin structure would provide less tortuous diffusion paths for ion migration. Such resins are termed *macroporous* and are made by employing organic solvents at the polymerization stage which are either compatible with (*sol-method*) or incompatible with (*nonsol-method*) the growing copolymer.

Sol-Method

If the initial polymerization mixture of monomer plus crosslinking agent contains a diluent which is a solvent (swelling agent) for the copolymer, for example methylbenzene, and if the fraction of crosslinking agent is high, then the solvent is not homogeneously distributed throughout the copolymer but occurs in localized regions bounded by densely crosslinked hydrocarbon chains. Upon removing the solvent by distillation the regions once occupied by solvent become distinct pores which are prevented from collapsing because of the rigidity imposed by the high degree of crosslinking.

Such a structure is termed *macroporous* with a typical average pore diameter of about 150 nm and a pore size range from several tens to several hundred nanometres. By comparison a gel resin is characterized by an 'apparent porosity' of no greater than about 4 nm which represents the average distance of separation of polymer chains. This difference in structural characteristics of gel and macroporous resins is clearly evident when comparing Figures 3.2a and 3.2b.

Nonsol-Method

Alternatively, the onset of macroporosity can be achieved by employing a diluent which does not solvate the copolymer, for example heptane. This is an example of the *nonsol* route and the porosity arises from the diluent precipitating out the copolymer containing localized pockets of solvent. Porosity is immediately apparent upon distilling off the solvent and the resin takes on a sintered appearance. In all the macroporous ion exchange resins the matrix is very heterogeneous in that it may be likened to a rigid pore (sponge like) structure supporting a variously crosslinked microporous gel matrix as shown schematically in Figure 3.1c. Both gel and macroporous structures are established for anion and cation exchangers of either styrenic or acrylic matrix and for all common classes of functionality, namely weak, strong, and ion selective.

ACRYLIC ANION EXCHANGE RESINS

In the case of the weakly acidic cation exchange resins the acrylic matrix arises from propenoic esters, nitriles, or alkyl substituted propenoic acids; this being the most obvious monomer by which to generate the carboxylic acid functional group. Where anion exchange resins are concerned the reasons behind wishing to establish an acrylic matrix were somewhat different. Excluding the fraction of crosslinking agent, the 'equivalent weights' of the unit styrenic and acrylic anion exchange resin structures are not vastly different. Thus given no great differences in intrinsic ion exchange capacity or functionality any differences in their behaviour must be a feature of their different skeletal structures.

This proved to be particularly true where the exchange of large complex organic anions were concerned. The affinity of such species for a resin is influenced not only by the ion charge but also by the structure of the ion and its size (ionic mass). The larger the organic anion the greater seems the affinity for an anion exchange resin due to van der Waals type intermolecular forces between the hydrocarbon structures of the organic ion and resin. A theoretical explanation of such behaviour is given in Chapter 5 when discussing origins of selectivity, but it suffices for now that providing the ion is not so large as to be excluded from the resin structure a general rule describing this type of intermolecular attraction is that 'like attracts like'. Therefore where fouling by large organic ions of *aromatic* structure is concerned van der Waals binding should be less pronounced if the skeletal structure of the resin is *aliphatic*.

This simple argument suggests that a gel or macroporous *acrylic* anion exchange resin should demonstrate a better resistance to irreversible fouling by high molecular weight aromatic organic anions compared with gel styrenic resins. In the case of humic or fulvic acids the above argument seems to be valid, and to such an extent that in water treatment should the problem of organic fouling persist the selection of acrylic anion exchange resins is almost always a considered option, should alternative resins foul irreversibly. A note of caution however: some site experiences can be cited which violate the above described argument especially where organic pollutants of industrial origin are concerned. In the absence of experience, resin selection with respect to organic fouling potential may often rely on site trials.

SOLVENT MODIFIED RESIN ADSORBENTS

Although not strictly within the context of ion exchange, mention should be made of the non-functional resin copolymers that, in recent times, have found application in the recovery and purification of pharmaceutical products such as vitamins and antibiotics. Another application of increasing importance is the removal of traces of noxious aromatic or aliphatic organic contaminants from polluted waters and waste effluents. In all cases the macroporous structure and the choice of aromatic (styrenic) or aliphatic (acrylic) matrix are important selection criteria together with the nature of the adsorbate. In some ways a comparison can be made with activated carbon sorption. In both cases sorption is usually less pronounced with increasing polarity of the adsorbate, and like carbon, the macroporous resin adsorbents possess a significant porosity as reflected by their internal surface areas of between about $100 \text{ m}^2\text{g}^{-1}$ and $900 \text{ m}^2\text{g}^{-1}$.

Unlike carbon however, the polymer adsorbents can be eluted (regenerated) with polar solvents such as methanol, propanone (acetone), electrolytes, or even water.

FURTHER READING

T. R. E. Kressman, 'Properties of Some Modified Polymer Networks and Derived Ion Exchangers', in 'Ion Exchange in the Process Industries', Society of Chemical Industry, London, 1970, p. 3.

R. Kunin, 'Pore Structure of Macroreticular Ion-Exchange Resins', in 'Ion Exchange in the Process Industries', Society of Chemical Industry, London, 1970, p. 10.

T. V. Arden, 'The Effect of Resin Structure', in 'Water Purification By Ion Exchange', Butterworths, London, 1968, Ch. 6, p. 99.

T. R. E. Kressman, *Effluent & Water Treat. J.*, 1966, **6**, 119.

Chapter 4

Properties and Characterization of Ion Exchange Resins

RESIN DESCRIPTION

The structure and ion exchange characteristics of resins available commercially commonly appear in the form of a data summary or 'specification' given in the product literature provided by all leading resin manufacturers. The terminology contained in such technical bulletins is listed in Table 4.1 and comprises a fairly thorough physical and chemical description of any resin. Typically, an ion exchange resin is described as being '*weak* or *strong*', '*acidic* or *basic*',

Table 4.1 *Classification of terms employed to describe ion exchange resins*

General Classification	
Chemical	*Physical*
matrix (polymer structure)	appearance (physical form)
crosslinking (% DVB)	particle size
functional group	uniformity coefficient
ionic form (as supplied)	grading
water content	density
ion exchange capacity	shipping weight
salt splitting capacity	percent whole beads
reversible swelling	sphericity
irreversible swelling	
pH range	
chemical stability	
thermal stability	

and '*cationic* or *anionic*'. The last classification is self-evident in referring to the charge on the counter-ion concerned, but the other terms in Table 4.1 require further explanation. In many respects ion exchange resins are solid phase equivalents of conventional aqueous electrolyte solutions as evident from the following considerations.

Firstly, ion exchange resins when hydrated generally dissociate to yield equivalent amounts of oppositely charged ions. Secondly, as with conventional aqueous acid or alkali solutions, resins in their acid or base forms may be neutralized to give the appropriate salt form. Finally, the degree of dissociation can be expressed in the form of an apparent equilibrium constant (or pK value) which defines the electrolyte '*strength*' of the exchanger and is usually derived from a theoretical treatment of pH titration curves.*

Strong Acid–Base Functionality

The terms *strong acid* and *strong base* may be defined in their simplest conventional electrolyte chemistry sense meaning that a strong cation resin in the acid form, or a strong anion resin in the base form, dissociates to give free hydrogen ions (H^+) and hydroxide (OH^-) ions respectively. The term *strong* has nothing at all to do with the physical strength of the resin, but instead derives from the Arrhenius Theory of electrolyte strength meaning complete dissociation of the functional group in whatever ionic form and at any pH. However, unlike ordinary aqueous strong electrolytes the dissociation of the functional group (acid, base, or salt form) *cannot* be detected in the external phase unless ion exchange occurs through the presence of other external counter-ions.

Equations 4.1–4.6 illustrate the fundamental conceptual steps involved in describing the dissociation characteristics of strong cation (sulfonic acid) and strong anion (quaternary ammonium hydroxide) exchangers, where the bar notation represents the resin phase. Once understood, dissociation in the resin phase is implied where appropriate, as illustrated by equations 4.7 and 4.8.

Strong Acid Cation ($R \equiv$ copolymer matrix)

$$\overline{RSO_3H} \rightarrow \overline{RSO_3H} \quad (4.1)$$
$$\text{anhydrous} \quad \text{no dissociation}$$

---

* The pK value of an acid (pK_a) or base (pK_b) is defined as the negative logarithm to base ten of the equilibrium dissociation constant K, whereby the more largely positive the pK value the more weak, or less pronounced, the dissociation of the acid ($R^- H^+$) or base ($R^+ OH^-$).

$$\overline{RSO_3H}_{\text{anhydrous}} + H_2O \xrightleftharpoons{\text{hydration}} \overline{RSO_3^- + H^+}_{\text{strong dissociation}} \quad (4.2)$$

$$\overline{RSO_3^- + H^+}_{\text{hydrated}} + A^+_{aq} \xrightleftharpoons{\text{ion exchange}} \overline{RSO_3^- + A^+} + H^+_{aq} \quad (4.3)$$

Strong Base Anion ($R \equiv$ copolymer matrix)

$$\overline{RCH_2N(CH_3)_3{}^+OH^-}_{\text{anhydrous}} \rightarrow \overline{RCH_2N(CH_3)_3{}^+OH^-}_{\text{no dissociation}} \quad (4.4)$$

$$\overline{RCH_2N(CH_3)_3{}^+OH^-}_{\text{anhydrous}} + H_2O \xrightleftharpoons{\text{hydration}} \overline{RCH_2N(CH_3)_3{}^+ + OH^-}_{\text{strong dissociation}} \quad (4.5)$$

$$\overline{RCH_2N(CH_3)_3{}^+ + OH^-}_{\text{hydrated}} + B^-_{aq} \xrightleftharpoons{\text{ion exchange}}$$
$$\overline{RCH_2N(CH_3)_3{}^+ + B^-} + OH^-_{aq} \quad (4.6)$$

Just as aqueous strong acids and strong bases may be neutralized to give a salt plus water, so is the case with strongly acidic or basic ion exchangers as illustrated by equations 4.7 and 4.8, where R represents the fixed polymeric anion or cation.

$$RH + NaOH \xrightarrow{\text{neutralization}} RNa + H_2O \quad (4.7)$$
$$ROH + HCl \xrightarrow{\text{neutralization}} RCl + H_2O \quad (4.8)$$

BOX 4.1 Experiment to Demonstrate the Dissociation Properties of Strong Functional Groups

Strongly functional cation and anion exchange resins when hydrated (swollen) dissociate (ionize) giving an internal electrolyte which is undetectable externally unless ion exchange occurs. This fundamental property of ion exchange resins is easily demonstrated using coloured acid–base indicators.

Requirements

strong acid cation exchange resin (H form)
strong base anion exchange resin (OH form)
two small funnels
two Petri dishes
two conical flasks (250 cm^3)
0.001 Molar hydrochloric acid
0.001 Molar sodium hydroxide
0.1 Molar sodium chloride
BDH 4.5 indicator solution
phenolphthalein indicator solution

filter papers
pipette droppers

Procedure

1. Indicator Reactions

Using the dilute acid and alkali solutions demonstrate the indicator reactions:

BDH 4.5, acidic – RED/YELLOW
non-acidic – GREY/BLUE

PHENOLPHTHALEIN, caustic (alkaline) – PINK
non-caustic, – COLOURLESS

2. Resin Only Reactions

a) SAC Resin (H Form). Place a nylon wool or foam plug at the base of a funnel resting in a conical flask. Transfer about 2–3 g of the resin to the funnel and rinse with demineralized water until fresh aliquots of the rinses give a non-acidic reaction to BDH 4.5 indicator. Place a small amount of the rinsed resin on a filter paper and moisten the resin with a few drops of BDH 4.5 indicator. The resin will turn red showing the presence of internal acidic dissociated hydrogen ions. Now cover the base of a Petri dish with demineralized water, add a fresh small quantity of rinsed resin, and a few drops of BDH 4.5 indicator. The non-acidic reaction shows hydrogen ions not to be present in the external water.

b) SBA Resin (OH Form). Similarly, the anion resin is rinsed until the rinses are non-caustic to phenolphthalein whereupon a small quantity of the resin will turn deep pink upon contacting phenolphthalein indicator – showing the presence of internal dissociated hydroxide ions. Equilibration of fresh resin with demineralized water shows no external caustic reaction to phenolphthalein indicator.

3. Exchange Reactions

These experiments are based on techniques described by Dorfner[1] and are particularly suited to demonstration using an overhead projector.

a) SAC Resin (H Form). Cover the bottom of a Petri dish with 0.1 Molar sodium chloride solution. Add a few drops of BDH 4.5 indicator to give a discernible blue colour. If the colour is too neutral (grey), add one or two drops of the 0.001 M sodium hydroxide solution. Now scatter a few beads of the rinsed cation resin across the surface of the salt solution. After a short while a yellow zone will form around the vicinity of the resin beads as hydrogen ions in the resin are exchanged for sodium ions in the external electrolyte producing dilute hydrochloric acid.

b) SBA Resin (OH Form). As described above, except that rinsed anion resin beads are scattered across the 0.001 M NaCl solution containing a few drops of phenolphthalein indicator. Ion exchange produces dilute sodium hydroxide shown by a deep pink zone spreading around the resin beads.

TIME REQUIRED – 30 minutes

Note: The appropriate ionic forms of the cation and anion resins may be prepared initially by passing 50 cm^3 of (a) 10% w/v HCl or (b) 10% w/v NaOH respectively slowly through 2–3 g amounts of resin contained in the funnel, followed by rinsing with demineralized water until rinses are negative to the appropriate indicator reaction. For procedure 3 an indicator which reacts at or around neutral pH, for example bromothymol blue, is particularly suitable for both demonstrations.

[1]K. Dorfner, 'Laboratory Experiments and Education in Ion Exchange', in 'Ion Exchangers', ed. K. Dorfner, Walter de Gruyter, Berlin and New York, 1991, Ch. 1.2, p. 437.

Weak Acid Cation Functionality

This class of cation exchange resin is typified by the carboxylic acid functional group which in its acid form is only very weakly ionized (pK_a approx. 4–6). The degree to which dissociation occurs is very pH-dependent, increasing with increasing external pH. Ion exchange of neutral salts by the acid form of the resin must, by definition, yield free hydrogen ions which would immediately displace the exchange equilibrium in the direction of functional group association. In other words, the great preference of the resin for hydrogen ions prevents any significant exchange of neutral cations. It is said that such a resin has only a limited salt splitting capacity. Neutralization with strong alkali solutions (high pH), however, will proceed to give the salt form, which being the salt of a weak acid–strong base is highly dissociated, and in this form could be correctly termed a strong exchanger.

Therefore it is important to understand that the weak/strong terminology is generic in referring to the parent acid form of the resin. Equations 4.9–4.13 illustrate the dissociation and ion exchange behaviour of ideal weakly acidic cation exchange resins where a short equilibrium arrow denotes an unfavourable direction of exchange.

$$\overline{RCOOH}_{\text{anhydrous}} \to \overline{RCOOH}_{\text{no dissociation}} \quad (4.9)$$

$$\overline{RCOOH}_{\text{anhydrous}} + H_2O \xrightleftharpoons{\text{hydration}} \overline{RCOO^- + H^+}_{\text{weak dissociation}} \quad (4.10)$$

$$\overline{RCOO^- + H^+} + A^+_{aq} \xrightleftharpoons{\text{ion exchange}} \overline{RCOO^- + A^+} + H^+_{aq} \quad (4.11)$$

$$\overline{RCOOH} + NaOH \xrightarrow{\text{neutralization}} \overline{RCOO^- + Na^+} + H_2O \quad (4.12)$$

$$\overline{RCOO^- + Na^+} + A^+_{aq} \xrightleftharpoons[(\text{pH} > \sim 4)]{\text{ion exchange}} \overline{RCOO^- + A^+} + Na^+_{aq} \quad (4.13)$$

This idealized description of weak acid resin characteristics suggests that all groups are identical regarding their dissociation strength, but this is in fact untrue. The reason for this is that for a given type of weak acid resin the heterogeneous distribution of functional sites gives rise to variations in charge density which results in there being significant variations in the value of the acid dissociation constant. This has a bearing upon the choice of weak acid exchanger for an intended application between a straight carboxylic acid resin [—RC(H)COOH] or, for example, the methylpropenoic acid analogue [methacrylic acid, —RC(CH$_3$)COOH] which possesses a weaker acid dissociation strength by virtue of the positive (electron repelling) inductive effect of the methyl group. Therefore an alkyl substituted carboxylic acid resin would be preferred for an application which relied upon the exchanger having the weakest obtainable acid strength, *i.e.* a resin possessing the highest pK_a value.

Weak Base Anion Functionality

The dissociation and subsequent ion exchange properties of weakly basic anion exchange resins are somewhat difficult to appreciate at first sight as no obvious ionization or dissociation path is evident from the group structure. It is initially instructive, though not actually valid, to propose ionization of the free amine group according to the Lewis Theory of acid–base reactions. Consider, for example, a tertiary weak base resin, RCH$_2$N(CH$_3$)$_2$. In the anhydrous state the amine group remains in the undissociated *free base form*. Upon solvation by equilibration with water it could be envisaged that protonation of the amine through the donor lone pair of electrons on the nitrogen atom neutralizes the conjugate acid (water) to give the ionized hydroxide form as depicted by equation 4.14.

$$\text{RCH}_2\text{N(CH}_3)_2 \underset{\text{anhydrous}}{} + \text{H}_2\text{O} \xrightarrow{\text{hydration}} \underset{\text{weak dissociation}}{\overline{\text{RCH}_2\text{NH(CH}_3)_2^+ + \text{OH}^-}} \quad (4.14)$$

Dissociation of the 'hydroxide' form is very weak since any significant concentration of hydroxide ions would immediately convert the resin back to the undissociated free base form. For the same reason ion exchange with a neutral anion as represented by equation 4.15 cannot substantially occur because of the implied liberation of free hydroxide ions, and an immediate reversal of exchange. In other words the hydroxide ion is so greatly preferred by the resin that ion exchange in strong alkali solutions is totally unfavourable, and the exchanger remains in the free base form (cf. weak acid resins).

$$\overline{\text{RCH}_2\text{NH(CH}_3)_2^+ + \text{OH}^-} + \text{B}^-_{aq} \xrightleftharpoons{\text{ion exchange}}$$
$$\overline{\text{RCH}_2\text{NH(CH}_3)_2^+ + \text{B}^-} + \text{OH}^-_{aq} \quad (4.15)$$

Reaction with strong acid solutions, however, will proceed since the reaction is one of neutralization to give the acid salt form as shown by equation 4.16.

$$\overline{\text{RCH}_2\text{NH(CH}_3)_2^+ + \text{OH}^-} + \text{HCl} \xrightarrow{\text{neutralization}}$$
$$\underset{\text{strong dissociation}}{\overline{\text{RCH}_2\text{NH(CH}_3)_2^+ + \text{Cl}^-}} + \text{H}_2\text{O} \quad (4.16)$$

Finally, as equation 4.17 shows, the salt form of the resin may ion exchange with other anions in the external solution providing the pH is sufficiently low to sustain the protonated state of amine nitrogen atom.

$$\overline{\text{RCH}_2\text{NH(CH}_3)_2^+ + \text{Cl}^-} + \text{NO}_3^-{}_{aq} \underset{(\text{pH} < \sim 9)}{\xrightleftharpoons{\text{ion exchange}}}$$
$$\overline{\text{RCH}_2\text{NH(CH}_3)_2^+ + \text{NO}_3^-} + \text{Cl}^-_{aq} \quad (4.17)$$

In conclusion, weakly basic exchangers in the free base form are only usefully functional at low pH when the hydrogen ion concentration is sufficiently high to protonate the resin. Therefore although conceptually useful, it is somewhat fallacious to propose the initially ionized 'hydroxide' form to explain the behaviour of weak base functional groups. Instead their reaction is better regarded as initially one of acid salt formation by a direct addition reaction and then where appropriate subsequent ion exchange, as depicted by equations 4.18–4.20.

$$\overline{\text{RN(CH}_3)_2} + \text{HCl} \xrightarrow{\text{acid addition}} \overline{\text{RNH(CH}_3)_2{}^+\text{Cl}^-} \quad (4.18)$$
<small>free base form   acid salt form</small>

$$2\,\overline{\text{RNH(CH}_3)_2{}^+\text{Cl}^-} + \text{SO}_4{}^{2-} \underset{(\text{pH} < \sim 9)}{\overset{\text{ion exchange}}{\rightleftharpoons}} \overline{[\text{RNH(CH}_3)_2{}^+]_2\text{SO}_4{}^{2-}} + 2\,\text{Cl}^- \quad (4.19)$$

$$\overline{\text{RNH(CH}_3)_2{}^+\text{Cl}^-} + \text{NaOH} \xrightarrow{\text{neutralization}} \overline{\text{RN(CH}_3)_2} + \text{NaCl} + \text{H}_2\text{O} \quad (4.20)$$

Weak acids such as carbonic acid (H_2CO_3) and 'silicic acid' (H_2SiO_3) are not sufficiently dissociated (strong) to protonate the weakly basic amine grouping, and therefore are not sorbed by a weak base ion exchange resin. This property is manipulated to great effect in such applications as water treatment by ion exchange.

The previous considerations now provide a basis for describing the principal chemical and physical characteristics of many commonly used ion exchange resins, as defined in the product technical data sheets available from all major resin manufacturers. Some selected values typical of such data are given in Tables 4.2 and 4.3 which along with the following discussion establishes a basis for optimum resin selection.

BOX 4.2 Experiment to Illustrate the Difference in Properties of Weak and Strong Functional Groups

Briefly, resins are of two kinds: insoluble organic acids are used for cation exchange and insoluble organic bases for anion exchange.

The cation exchangers are either sulfonic or carboxylic acids, the difference being that sulfonic acids are always ionized and therefore may be used at any pH, whilst carboxylic acids are only ionized at about pH 7 or above and can only be used for alkali treatment.

$$RSO_3H + NaCl \rightarrow RSO_3.Na + HCl$$
$$RSO_3H + NaOH \rightarrow RSO_3.Na + HOH$$
$$RCOOH + NaCl \rightarrow \text{no reaction}$$
$$RCOOH + NaOH \rightarrow RCOO.Na + HOH$$

The anion exchange resins are quaternary or tertiary amines. Quaternary amines are always ionized and their reaction is independent of pH, while a tertiary amine absorbs acids by a single addition and can therefore only operate at below pH 7.

$$RNX_3.OH + NaCl \rightarrow RNX_3.Cl + NaOH$$
$$RNX_3.OH + HCl \rightarrow RNX_3.Cl + HOH$$
$$RNX_2 + NaCl \rightarrow \text{no reaction}$$
$$RNX_2 + HCl \rightarrow RNX_2H.Cl$$
$$(X \equiv \text{alkyl group}, e.g. -CH_3)$$

Requirement

Glass column 6 mm diameter (approx.)
5 ml strong cation resin, SAC
5 ml weak cation resin, WAC
5 ml strong anion resin, SBA
5 ml weak anion resin, WBA
1 M hydrochloric acid
0.1 N NaOH/0.1 N NaCl solution
0.1 N HCl/0.1 N NaCl solution
0.1 N NaOH
0.1 N HCl
0.1 N $AgNO_3$
100 ml volumetric flask
Phenolphthalein indicator
Potassium chromate indicator

Procedure – Cation Resins

1. Put the resins into the glass columns.
2. Regenerate each with 6 BV 1 N HCl and rinse with 6 BV deionized water at 1 BV/3 minutes.
3. Pipette 25 ml of 0.1 N NaOH/0.1 N NaCl solution onto each resin and pass it through at 1 BV/10 minutes, collecting the eluate in a 100 ml volumetric flask. Rinse the resins with deionized water and make volumetric flasks up to the mark.
4. Titrate 20 ml aliquots with 0.1 N NaOH using phenolphthalein indicator, then titrate the neutralized solution with 0.1 N $AgNO_3$ using potassium chromate indicator.

Results

Both resins exchange sodium from sodium hydroxide for hydrogen to form water but only the strong resin gives an acidic eluate due to the exchange of hydrogen for sodium from the sodium chloride.

TIME REQUIRED – 2 hours

Procedure – Anion Resins

1. Put resins into glass columns.
2. Regenerate the resins with 6 BV 1 N NaOH and rinse with 6 BV deionized water at 1 BV/3 minutes.
3. Pipette 25 ml of 0.1 N HCl/0.1 N NaCl solution on to each resin and pass it through at 1 BV/10 minutes, collecting the eluate in a 100 ml volumetric flask. Rinse with deionized water and make up to the mark.
4. Titrate 20 ml aliquots with 0.1 N HCl and then with 0.1 N $AgNO_3$ when neutralized.

Results

Both resins exchange chloride ions from hydrochloric acid for hydroxide ions to form water, but only the strong resin gives an alkaline eluate due to the exchange of hydroxide ions for chloride from the sodium chloride.

TIME REQUIRED – 2 hours

CHEMICAL SPECIFICATION

Matrix

The common choice is between 'styrene–divinylbenzene' or 'acrylic–divinylbenzene' copolymer. In the case of cation exchange resins selection is easily made since the acrylic products are weakly acidic whilst the styrenic resins are strongly acidic. Therefore for cation exchange the choice of copolymer is primarily decided by the process application and operating pH. The situation is very different with anion exchange resins since the two types of matrix pertain to products of both weak and strong functionality. Where anion exchange resins are concerned the choice between an acrylic resin and its styrenic equivalent is often made on considerations of operating exchange capacity, physical strength, and fouling resistance to complex high molecular weight organic anions.

Disregarding structural features (gel or macroporous) for the time being, the acrylic matrix is particularly tough being more elastic than the more rigid styrene-based copolymer. However the elastic resilience of the acrylic matrix could be of concern where columns of resin operate under a high net compression force (hydraulic pressure drop) since this gives rise to resin bead compression and bed compaction resulting in impeded flows and poor liquid phase distribution (channelling). It must be stressed that in relation to mechanical strength there are no infallible *rules* as such, but rather *guidelines* for resin selection which very much depend upon the nature of the application and the associated equipment engineering design. The type of copolymer matrix is not usually the single most important selection criterion with regard to physical strength since both resin structure and degree of crosslinking contribute greatly in this respect.

Structure

Here, the choice is between *gel* and *macroporous* materials, all other considerations assumed equal. The polymer structure of the resin influences the mechanical strength, swelling characteristics, ion exchange equilibria, and exchange kinetics properties of all resins. Macroporous copolymers, being highly crosslinked, are generally tougher than their gel equivalents and are more resistant to physical breakdown through mechanical forces, osmotic volume changes, and chemical degradation of crosslinking through the action of oxidizing agents.

Some product data bulletins discriminate between resin structures by way of reporting higher breaking weights (gram per bead) for beads of macroporous resins compared with gel equivalents, but care must be exercised in the interpretation of such data since the forces required to fracture a 'strain free' gel resin bead and a macroporous equivalent may be quite comparable. Undeniably, statistical analysis of bead fracture studies yield useful information about structural fault propagation and matrix elasticity, but to take 'breaking weight' in isolation as a selection parameter for gel versus macroporous resins could be misleading since the basis of the comparison is not usually disclosed.

Macroporous resins, being highly crosslinked, possess quite a heterogeneous distribution of structurally dense and tortuous regions of high charge density, and it is for this reason that the affinity of a macroporous resin for a given inorganic ion is usually greater than that for a gel resin, and sometimes the rate of exchange can be discernibly slower for macroporous resins when compared with gel equivalents. But where exchange of large, high molecular weight species are concerned the macroporous property becomes important in providing an easier diffusion path for the uptake and subsequent release of such species. Finally, because macroporous resins possess real pores, the number of functional groups per unit dry weight of matrix is usually less than that for an equivalent gel product as reflected by their slightly lower dry weight capacities—see Table 4.2 and 4.3.

Table 4.2 Selected properties of some typical cation exchange resins

Resin type	Matrix	Structure	Functional group	Ionic form	WVC (keq m^{-3})	pH range	Thermal stability (°C)	Reversible swelling (%)
Strong acid	styrene–DVB	gel	—SO$_3^-$	Na H	2.0 1.8	0–14	120	Na → H; 7
		macroporous	—SO$_3^-$	Na H	1.8 1.6	0–14	120	Na → H; 3–5
Weak acid	acrylic–DVB	gel	—COO$^-$	H	4.2	4–14	120	H → Na; 70–100 H → Ca; 20
		macroporous	—COO$^-$	H	3.0	4–14	120	H → Na; 50 H → Ca; 15

(Values typical of standard resins. WVC = wet volume capacity)

Table 4.3 Selected properties of some typical anion exchange resins

Resin type	Matrix	Structure	Functional group	Ionic form	WVC (keq m^{-3})	pH range	Thermal stability (°C)	Reversible swelling (%)
Strong Base (Type 1)	styrene–DVB	gel	—N(CH$_3$)$_3$$^+$	Cl	1.3	0–14	80 (Cl) / 40 (OH)	Cl → OH; 20
		macroporous			1.15			
Strong Base (Type 2)	styrene–DVB	gel	—N(CH$_3$)$_2$(CH$_2$CH$_2$OH)$^+$	Cl	1.3	0–14	60 (Cl) / 40 (OH)	Cl → OH; 15–20
		macroporous			1.15			
Strong Base (Type 1)	acrylic–DVB	gel	—N(CH$_3$)$_3$$^+$	Cl	1.25	0–14	75 (Cl) / 35 (OH)	Cl → OH; 10–15
		macroporous			1.2			
Weak Base	styrene–DVB	gel	—N(CH$_3$)$_2$	Cl	1.2	0–9	100	FB → Cl; 10
		macroporous		free base	1.25		100	FB → Cl; 15–20
		gel	polyamine	free base	1.9		100	FB → Cl; 10–15
Weak Base	acrylic–DVB	gel	—N(CH$_3$)$_2$	free base	1.6	0–9	60	FB → Cl; 15–20
		macroporous			1.0			
Mixed Base	acrylic–DVB	gel	—N(CH$_3$)$_3$$^+$ / —N(CH$_3$)$_2$	Cl / free base	1.3	0–14	35 (OH)	OH → Cl; 5

(Values typical of standard resins only. WVC = Wet Volume Capacity; FB = free base)

Crosslinking

Commercial divinylbenzene is not pure, but rather a mixture of various isomers plus some ethylstyrene. The percent by weight of this mixture making up the copolymerization reactants is termed the nominal percent crosslinking or percent DVB. Standard gel resins are produced with about 8% DVB, although the macroporous varieties may report from around 15% DVB to as high as 30% crosslinking depending upon whether made by the '*nonsol*' or '*sol*' route respectively.

Crosslinking provides the fundamental chemical bonding between adjacent polymer chains thus giving the resin its inherent physical strength. The degree of crosslinking also governs the extent of swelling of the dry ion exchange resin upon absorbing water. The more weakly crosslinked the resin the greater the swelling and water uptake. Gel resins are made ranging from very low (0.5% DVB) to around 25% to suit a range of applications, but for industrial uses a weakly crosslinked product would be too soft whilst an excessively crosslinked gel resin would be brittle, relatively slow to exchange ions, and would show unfavourable operational equilibrium properties.

Functional (or Ionogenic) Group

The nature of the fixed ion generically classifies the functionality of resins as follows:

1. *Strong acid cation* – sulfonate, $-SO_3^-$
2. *Weak acid cation* – carboxylate, $-COO^-$
3. *Strong base anion* – Types 1 and 2
   Type 1 – benzyltrimethylammonium, $-CH_2N(CH_3)_3^+$
   Type 2 – benzyldimethylethanolamine,
   $-CH_2N(CH_3)_2(CH_2CH_2OH)^+$

All common strong base anion exchange resins are 'quaternary ammonium' derivatives but Types 1 and 2 differ in their basicity. Type 2 resins are slightly less basic than their Type 1 equivalents which results in the former materials exhibiting a slightly higher affinity for hydroxide ions.

Ionic Form

By definition an ion exchange resin may be converted to virtually any counter-ion form, but commercially the most common ionic forms as manufactured are:

1. *Strong acid cation*
   Hydrogen form, —$SO_3^-H^+$
   Sodium form, —$SO_3^-Na^+$
2. *Weak acid cation*
   Hydrogen form, —COOH
3. *Strong base anion*
   Type 1, chloride form, —$CH_2N(CH_3)_3^+Cl^-$
   Type 2, chloride form, —$CH_2N(CH_3)_2(CH_2CH_2OH)^+Cl^-$
4. *Weak base anion* (tertiary amine)
   Free base form, —$CH_2N(CH_3)_2$
   Chloride form, —$CH_2NH(CH_3)_2^+Cl^-$

Most ionic forms may be prepared by passing a large excess of an appropriate acid, alkali, or salt solution at 1–2 g equiv l^{-1} concentration through a column of resin over 20–30 minutes (see Box 4.3). The ease of resin conversion generally increases with decreasing particle size, decreasing crosslinking, and decreasing charge of the ion being displaced.

BOX 4.3 Conversion of Resins to Standard or Alternative Ionic Forms

Introduction

Most simple ion forms of strongly functional ion exchange resins may be prepared by passing 5–10 bed volumes (BV) of a 10% w/v solution of the desired displacing or eluting ion through a *column* of resin at a rate of about 1 BV in 3 minutes. This is followed by a rinse with deionized water (5–10 BV) until testing negative for the ion concerned. Care must be taken to ensure that the conversion being carried out does not permit any precipitation reactions to occur.

Of particular interest is the conversion of hydroxide form strongly basic anion exchange resins to a salt form using polybasic acids. For example, with sulfuric acid initial neutralization of the acid by the resin promotes formation of the sulfate form, but as the converted resin contacts excess acid the hydrogensulfate form is produced. This mixed ionic equilibrium condition is avoided using a neutral salt to convert the resin, e.g. sodium sulfate. Not dissimilar considerations apply to preparing the hydrogencarbonate form of a strong base anion resin. This is not possible commencing with the resin in the hydroxide form since the carbonate form results and it is therefore necessary to proceed via the chloride form. Weakly functional cation and anion exchange resins can only be converted from their acid or free base forms under alkaline or acid conditions respectively. Once prepared, any resin in the desired ionic form may be filtered using a Buchner flask and stored in a damp condition.

When using a resin from new it is advisable to pre-condition the material by alternately cycling and rinsing the resin between acid and salt form (Na) for cation exchangers or hydroxide and salt form (Cl) for anion exchangers. This serves to osmotically cycle the resin and flush out residual impurities left over from the manufacturing process.

Measurement of Resin Volume, *i.e.* Bed Volume

A volume of wet resin should be measured in a measuring cylinder of suitable size using sufficient water to provide at least 2–3 cm depth of supernatant water. If the side of the cylinder is tapped with a rubber bung the resin will be seen to settle down, and will have reached a minimum volume after 3–4 minutes. This volume, read from the graduations on the side of the cylinder, is usually termed the minimum vibrated or tapped down volume, and is used for the volumes of resin described in the experiments.

Note: It is often convenient to provide the bung with a handle by mounting it on a rod.

Bed Volume

As resin columns of different sizes may be used, it is frequently convenient for volumes of regenerants and other liquors to be related to the volume of resin used. Thus the term *bed volume* (BV) is used and it refers to the volume of resin as measured by 'tapping down'. For example, if 25 ml of resin are used, then 3 bed volumes of regenerant would be 75 ml.

Setting-up an Ion Exchange Column

A glass column similar to one of those shown in the Figure is selected or prepared. Essentially the column consists of a tube with a narrow outlet at the lower end through which liquor can flow. To control the rate of flow a glass tap, or rubber tubing with a screw clip, should be fitted in the outlet. Glass wool, sand, a sintered disc, or plastic foam plug should be inserted in the lower end of the main tube as a supporting medium for the resin. In acid or alkaline solutions above a concentration of 2 N, use a suitable inert support.

The tube is filled with water to a depth of 5–7 cm and a slurry of the resin poured into the column. The tube length should be selected so that the volume of water above the resin bed is at least half the volume of resin used. Deionized water is used to prepare the resin column. It is important not to allow the water level in the column to run below the top of the resin bed level or to allow the column to run dry. Air pockets are formed in the resin bed and this prevents uniform flow of liquid through the column. If the column runs

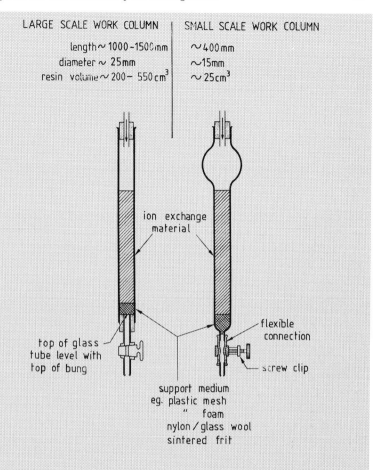

dry, water should be passed up through the base of the column until the air has been displaced. This procedure is known as 'backwashing'.

Deionized Water

Wherever this is specified, distilled water may be used as an alternative.

Water Content

Water content or swelling water as it is sometimes called can be looked upon as a measure of the 'water of hydration' of an ion exchange resin. Water enters a dry resin thereby hydrating the fixed ions and counter-ions and at the same time causing the resin to swell against the restraining action of the crosslinks. Eventually an osmotic equilibrium is reached whereby the internal swelling pressure within the resin opposes any further water uptake (see Chapter 5, 'Swelling Phenomena and the Sorption of Solvents'). The swelling of a resin as measured by its water content is inversely related to the degree of crosslinking and is a most important structural characteristic. Should the water content of a given resin, as measured in a standard form, increase significantly this is indicative of a reduction in crosslinking through some kind of attack on the resin which should prompt immediate investigation.

The *water content* of a resin is usually expressed as a percent of its swollen weight, but some authorities prefer to express water content as the weight of water taken up by 1 gram of initially dry resin, in which case it is termed *water regain*. If the water content is W percent and the water regain WR (g H_2O per dry gram), the relation between the two values is:

$$W = 100\left(\frac{WR}{WR + 1}\right)\% \qquad (4.21)$$

The water content is easily determined by weighing a quantity of resin before and after drying at 105 °C, but the most precise determinations incorporate a centrifuging step in order to correct for interstitial water adhering to the surface of the beads (see Box 4.4). Sometimes it may be desirable to employ drying methods at reduced temperatures under vacuum to minimize errors arising from possible thermal degradation, a consideration that could sometimes apply to anion exchange resins.

BOX 4.4 Simplified Measurement of Water Regain and Voids Volume

A sample of wet resin contains water within the beads and also between them. The water regain measures the water within the beads, whilst the voids volume is a measure of the interstitial water. The water regain of a resin is one of its fundamental characteristics and varies inversely with the crosslinking of the resin.

Requirements

50 ml measuring cylinder
25 ml (approx.) of SAC resin (H form) or SBA resin (Cl form)

2 N hydrochloric acid
Wetting agents: 0.2% Teepol for cation resins
  0.2% Cirrasol for anion resins
12 mm diameter column
Screened methyl orange indicator
Dry 100 ml Buchner flask
12 mm diameter column (dry)

Procedure

1. The resin must be fully converted to the standard form: H for cation, and Cl for anion resins.
2. Place about 25 ml of resin in the column and regenerate with 200 ml of 2 N HCl at 1 BV/2 minutes.
3. When the acid has drained to bed level, rinse with deionized water at 1 BV/2 minutes until the eluate is just alkaline to screened methyl orange indicator.
4. Measure 20–25 ml of the regenerated resin in the 50 ml cylinder and add about 25 ml of supernatant water. Tap the resin down and record the volumes of resin and supernatant water.
5. Add a few drops of the appropriate wetting agent and transfer the resin and water to the dry column, which should be connected to the Buchner flask with a bung.
6. Apply gentle suction to the flask and collect the filtrate. Use the filtrate to rinse all the resin into the column, and when all is transferred suck the resin dry for 5 minutes.
7. Measure the volume of the filtrate; this gives the volume of supernatant water and voids.
8. Weigh accurately about 5 g of the moist resin, and dry (usually overnight) to a constant weight at 100 °C.

Voids Volume

$$\text{Volume of supernatant water} = x \text{ ml}$$

$$\text{Volume of resin and voids} = y \text{ ml}$$

$$\text{Volume of supernatant water and voids} = z \text{ ml}$$

Therefore

$$\text{Voids} = z - x \text{ ml}$$

$$\text{Voids volume} = \frac{z - x}{y} \times 100\% \text{ of the resin bed volume}$$

(usually between 35% and 40%)

Water Regain

Weight of wet resin = a g

Weight of dry resin = b g

$$\text{Water regain} = \frac{a-b}{b} \text{ g of water per g of dry resin}$$

$$\text{Water content} = \frac{(a-b)}{a} \times 100\%$$

TIME REQUIRED – 1 hour 45 minutes plus overnight drying

N.B. For more accurate determinations of water regain only, the voids water is not removed by suction but extracted by centrifuging.

pH Range

This is very much dependant upon the *strength* of the functional group as previously discussed, but the following guidelines apply:

1. *Strong acid cation*: any pH
2. *Weak acid cation*: > 4
3. *Strong base anion*: any pH
4. *Weak base anion (tertiary)*: < 9

Reversible Swelling

Reversible swelling refers to the observable and reversible volume changes that occur between one swollen ionic form and another for a given resin. The volume changes reflect the differing magnitude of resin–counter-ion interactions, percent crosslinking, and degree of hydration of the ions and matrix concerned. From a practical standpoint an awareness of resin volume changes between different initial and final ionic states is important when designing ion exchange plant for a specific duty. Most ion exchange processes involve cycling resins between different ionic forms and therefore the resin matrix has to be able to withstand the many osmotic cycles occurring over the projected practical lifetime of the resin. Also, it is not always appreciated that initially the swollen *ionized* forms of all resins shrink upon contact with an electrolyte, the degree of shrinking being greater the more concentrated the external solution. Subsequent equilibration with water (rinsing) swells the resin and establishes its final swollen volume.

Cycling resins between extremes of external ion concentration can lead to bead fracture due to *osmotic shock*, and if unavoidable, could well be a factor which decides in favour of macroporous over gel materials.

Irreversible Swelling

Irreversible swelling is a phenomenon principally observed with acrylic strong base anion exchange resins whereby upon undergoing their first few aqueous ion exchange cycles an irreversible expansion occurs of around 7–10% over and above the reversible volume changes which thereafter apply.

Chemical Stability

At the low sub-microanalytical level evidence exists for trace resin solubility through release of monomer and copolymer degradation products ('leachables'). An awareness of this is proving to be particularly important in such applications as '*ultrapure*' water production for the microelectronics, pharmaceutical, and power generation industries, but in no way does it lessen the indispensable role played by ion exchange in these technologies. At the macroscopic level the chemical stability of modern resins at normal ambient temperatures is excellent, being insoluble in all common organic solvents and electrolyte solutions.

Two principal exceptions are resin breakdown caused by sustained exposure to ionizing nuclear radiation and powerful chemical oxidizing agents such as nitric acid, chromic(VI) acid, chlorate(V) ions, halogens, and peroxy compounds. Even saturation levels of dissolved oxygen in the presence of transition metal cations may initiate chemical breakdown albeit only relatively slowly at ambient temperatures.

CAUTION: Under no circumstances should anion exchange resins be contacted with concentrated nitric acid since subsequent reactions of explosive violence have been known to occur.

Exposure to dilute nitric acid is allowed, but precautions should be taken to avoid any accidental increase in nitric acid concentration, together with a means of venting any possible build up of pressure. Operations with nitric acid of any concentration at elevated temperatures is to be avoided. Nitric acid hazards do not seem to be prevalent with cation exchange resins, but with high concentrations (greater than about 3 g equiv l^{-1}), oxidative degradation of crosslinking can occur.

Oxidative attack generally results in progressive decrosslinking of a resin, whereas most other instances of chemical degradation manifests itself either as a loss of exchange capacity or a change in the basicity of the functional group. All cation exchange resins, and weakly basic anion exchange resins are fairly stable with regard to capacity loss and their useful operating life is more influenced by mechanical stresses and the presence of foulants. The situation is somewhat different with strong base anion exchange resins since the quaternary ammonium functional group is inherently more unstable. An average strong base capacity loss of approximately 5% and 10% per year is not unusual for Type 1 and Type 2 materials respectively. The hydroxide form of a strong base resin is the most unstable form, and if working resins are to be stored for any considerable period they are best maintained wet and in their respective standard forms, namely: hydrogen or sodium for cation exchange resins and chloride for anion exchangers.

Ion Exchange Capacity

Exchange capacity is possibly the most important characteristic of an ion exchange material since it is a measure of its capability to carry out useful ion exchange work. Several definitions are employed depending upon the intended application of the data.

Dry Weight Capacity (DWC)

This is sometimes called the intrinsic or specific capacity and is defined as the total number of equivalents of exchangeable ion, in a stated form, per dry kilogram of the resin [milliequivalents (meq) per dry gram]. In the context of ion exchange equivalent mass is defined as the gram ion mass (or molar mass) per unit ion charge, and dry weight capacity is the prime capacity defining characteristic of a resin as manufactured. The standard ionic states are hydrogen form and chloride form for cation and anion exchange resins respectively, and to a fair approximation exchange capacity values can be predicted from the equivalent mass of the monomer characterizing the exchanger. For example, the empirical formula of the functional monomer for a gel styrenesulfonic acid resin in the hydrogen form may be written $C_8H_7.SO_3H$, with a Relative Molecular Mass (mass of 1 mole) of 184. Thus the unit equivalent mass is 184 g containing one equivalent of exchangeable hydrogen ions from which the anticipated exchange capacity is 1 equivalent per 184 g dry resin, i.e. 5.4 equiv. per kilogram ($eq\,kg^{-1}$) dry resin in the hydrogen form (5.4 milliequivalents per dry gram, $meq\,g^{-1}$).

However, the real structures incorporate divinylbenzene and are not homogeneous which results in the measured total exchange capacities being a little lower than those given by the afore-described simple model. Similar considerations when applied to anion exchange resins give values appreciably greater than their measured capacities owing to most commercial resins only being 70–85% chloromethylated.

The measurement of dry weight capacity in a standard ionic form is usually carried out by direct titration of the exchanged ion using either a weighed quantity of dried resin or alternatively a known mass of swollen resin whose water content is determined separately (see Box 4.5). The neutralization of ion exchange resins in their acid and base forms by addition of standard alkali or acid solutions respectively may be easily studied by pH titration. The titration curves obtained are similar to those found for conventional acid–base systems and are shown in Figure 4.1. The capacity of the resin is found from the points on the titration curve where the rate of change in pH with titrant addition is greatest. Dependant upon whether the reacting functional groups are strong (acid or base), weak acid, or weak base, the pH at complete neutralization will be neutral (pH = 7), alkaline (pH $\gg$ 7), or acidic (pH $\ll$ 7) respectively.

By way of example, a pH titration of a sulfonic acid resin (12% DVB) with sodium hydroxide solution gave the following results from which the dry weight capacity may be calculated:

1. mass of swollen resin 0.97 g (H form)
2. water content 35%
3. volume of sodium hydroxide solution required for neutralization (to pH 7) = 28.4 cm^3 = 0.0284 dm^3
4. concentration of sodium hydroxide solution = 0.1 $mol\,dm^{-3}$

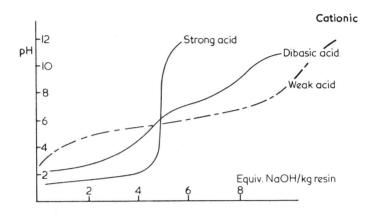

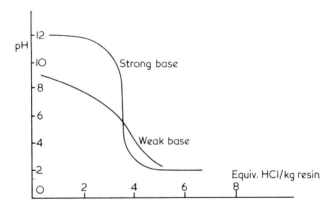

Figure 4.1 *pH titration curves of representative resin types*
(Reproduced from R. W. Grimshaw and C. E. Harland, 'Ion-Exchange: Introduction to Theory and Practice', The Chemical Society, London, 1975)

Therefore:

$$\text{Dry Weight Capacity} = \frac{0.0284 \times 0.1}{0.97 \times (1 - 0.35)} \, \text{eq g}^{-1}$$

$$\text{DWC} = 4.5 \text{ milliequivalents per dry gram}$$
$$(i.e.\ 4.5 \text{ meq g}^{-1}, \text{H form})$$

BOX 4.5 Determination of Dry Weight Capacity

1 Determination of the Capacity of a SAC Exchanger on a Dry Weight Basis

The total capacity of a SAC exchanger is determined by converting it to the hydrogen form, and then using a neutral sodium salt to displace the hydrogen ions which are titrated as free acidity.

Requirements

Stirrer
0.1 N sodium hydroxide
Analar sodium chloride
Screened methyl orange
Approx. 0.5 g of dry strong cation resin (H form)

Procedure

1. Weigh accurately about 0.5 g of resin and transfer it to a 100 ml beaker. Add about 25 ml of deionized water.
2. Add about 2 g of Analar sodium chloride and stir.
3. Titrate the mixture with 0.1 N NaOH using screened methyl orange indicator.
4. The end point may be reversed several times during stirring as more hydrogen ions are exchanged, but a steady end point should be obtained within a reasonable time. It is advisable to add the last portions of NaOH more slowly.

Let

$$\text{Weight dry resin} = W \text{ g}$$
$$\text{NaOH normality} = N$$
$$\text{Titre} = T \text{ ml}$$

Then

$$\text{Dry weight capacity} = \frac{TN}{W} \text{ meq/dry g}$$

TIME REQUIRED – 45 minutes

2. Determination of Total Acid Capacity of a WAC Exchanger on a Dry Weight Basis

Procedure

1. The resin is first converted to the standard H^+ form and dried to constant weight at 100 ± 5 °C.
2. About 0.5 g of the dry resin (W g) is accurately weighed into a conical flask containing 25 ml of mixed bed water.
3. To this is added 15 ml of 1 N NaOH measured by pipette.
4. The flask is then stoppered and left to stand for 4 hours with occasional shaking.
5. The contents of the flask are then back titrated with 0.5 N HCl using thymolphthalein indicator solution (t ml).
6. Capacity of $\dfrac{15\,N_1 - N_2 t}{W}$ meq g^{-1}

   Where: W = weight of dry resin (g)
   N_1 = normality of standard NaOH
   N_2 = normality of standard HCl
   t = titre of standard HCl (ml)

Note: Where a value of < 10.0 meq g^{-1} is obtained, the procedure should be repeated using 10 ml of sodium hydroxide solution.

TIME REQUIRED – 45 minutes

3. Determination of the Capacity of a SBA Exchanger on a Dry Weight Basis

The total capacity of a SBA exchanger is determined by converting it to the chloride form, and then using a neutral nitrate to displace the chloride ions which are then estimated.

Requirements

Stirrer
0.1 N silver nitrate
Analar sodium nitrate
Potassium chromate indicator
Approx. 0.5 g of dry strong anion resin (Cl form)

Procedure

1. Weigh accurately about 0.5 g of resin and transfer it to a 100 ml beaker. Add 25 ml of deionized water.
2. Add about 2 g of Analar sodium nitrate and stir.

3. Titrate the mixture with 0.1 N $AgNO_3$ using potassium chromate indicator.
4. The end point may be reversed several times during stirring as more chloride ions are exchanged, but a steady end point should be obtained within a reasonable time. It is advisable to add the last portions of $AgNO_3$ more slowly.

Let

$$\text{Weight dry resin} = W \text{ g}$$
$$AgNO_3 \text{ normality} = N$$
$$\text{Titre} = T \text{ ml}$$

Then

$$\text{Dry weight capacity} = \frac{TN}{W} \text{ meq/dry g}$$

TIME REQUIRED – 45 minutes

Special applications or experimental studies may require values for dry weight capacity in non-standard ionic forms. In this case the usual method is one of column elution of the swollen resin of known water content, and subsequent analysis of the collected eluate for the exchanged ion. Where simple inorganic ionic forms are concerned the dry weight capacity may be calculated from the measured value in the standard form.

For example, the dry weight capacity of a 12% crosslinked styrenesulfonic acid resin was found to be 4.53 equiv. per dry kilogram, H form. In other words, one kilogram of dry resin is made up of 995.47 g dry sulfonated matrix plus 4.53 g exchangeable hydrogen ions, or 1 equivalent per 219.8 g matrix. If all the hydrogen ions are exchanged for sodium ions one equivalent of exchangeable sodium ions are incorporated within a resin of total dry weight 219.8 g plus 23.0 g equals 242.8 g. Thus the dry weight capacity in the sodium form is given by 1/242.8 equiv. per dry gram or 4.12 eq kg^{-1} (Na form).

Of course, should complete ion conversion of the resin be prevented for any reason the calculated capacity will be too high and actual measurement must be undertaken. Table 4.4 shows a comparison of measured and calculated dry weight capacities for different ionic forms of styrenesulfonic acid resins of different crosslinking.

Table 4.4 *Comparison of measured and calculated dry weight capacities of sulfonic acid cation exchange resins*

Ionic form	Measured DWC (eq/dry kg) 1% DVB	12% DVB	Calculated DWC (eq/dry kg) 1% DVB	12% DVB
H^+	5.1	4.5	—	—
Na^+	4.7	4.2	4.6	4.1
Ca^{2+}	4.7	4.2	4.7	4.2

(DWC ≡ dry weight capacity)

Wet Volume Capacity (WVC)

Most laboratory and industrial applications of ion exchange employ resins in columns or cylindrical pressure vessels respectively. Hence a value for total exchange capacity on a bulk (apparent) volume basis is more meaningful. Elution methods are used in volume capacity determinations whereby the total number of equivalents of exchangeable ions per unit volume of the resin is measured and the result usually expressed as equivalents per litre of resin ($eq\, l_R^{-1}$). See Box 4.6.

BOX 4.6 Determination of Wet Volume Cation Capacity

The total capacity of a SAC resin is determined by converting it to the hydrogen form, and then using a neutral sodium salt to displace the hydrogen ions which are titrated as free acidity.

Requirements

Glass column, 12 mm diameter (approx.)
25 ml SAC resin (Na form)
2 N hydrochloric acid
5% sodium chloride solution
Screened methyl orange
0.1 N sodium hydroxide
250 ml volumetric flask

Procedure

1. Put the resin into the column and regenerate it with 200 ml of 2 N HCl at 1 BV/3 minutes.

2. When the acid has drained to bed level, rinse with deionized water at 1 BV/minute until the eluate is just alkaline to screened methyl orange.
3. Transfer the resin to a measuring cylinder and tap it down under water. Record the volume of resin. Carefully put the resin back into the column and drain to bed level.
4. Run 200 ml of 5% w/v NaCl through the bed at 1 BV/2 minutes. Collect the eluate in a 250 ml volumetric flask.
5. Run 50 ml of deionized water through the bed at 1 BV/2 minutes. Collect sufficient of the rinse to make up the original volume.
6. Titrate a suitable aliquot (*e.g.* 10 ml) with 0.1 N NaOH using screened methyl orange indicator.

Let

$$\text{Resin volume} = V \text{ ml}$$
$$\text{Aliquot of solution} = A \text{ ml}$$
$$\text{Titre} = T \text{ ml}$$
$$\text{NaOH normality} = N$$
$$\text{Capacity} = \frac{250 TN}{AV} \text{ eq/litre of hydrogen form resin}$$

TIME REQUIRED – 1 hour 45 minutes

Strong and Weak Functional Capacity

For resins described as strong or weak the reported exchange capacity might be expected to relate to one class of functional group only. However some products, particularly anion exchange resins, incorporate a proportion of functional groups whose basicity is opposite to that implied by its nominal description. This occurs for several reasons:

1. Side reactions occurring during resin synthesis.
2. A deliberate inclusion of opposing basicity groups in order to minimize or optimize volume changes.
3. Deliberate manufacture of a multifunctional resin.
4. Degradation of strong to weak functionality during the working lifetime of the resin. This is particularly relevant to *Type 2* strong base anion exchange resins which whilst they maintain a useful total exchange capacity do degrade to weak base at a significant rate. Therefore monitoring such degradation with time is important in assessing when such a resin may be no longer expected to function economically as a strong base exchanger. This is

achieved by carrying out regular capacity checks using methods which differentiate between strong base and weakly basic groups (see Box 4.7).

BOX 4.7 Determination of Strong and Weak Base Wet Volume Capacity of an Anion Exchange Resin

The total capacity of the resin in the chloride form is estimated, and followed by the determination of the chloride capacity of the strong groups only. The capacity of the weak groups is found by difference.

Requirements

Glass column 12 mm diameter (approx.)
25 ml SBA or WBA resin (Cl form)
0.5 N hydrochloric acid
1% sodium nitrate
2.5% sodium chloride
0.1 N silver nitrate
2.5 N ammonium hydroxide solution
1 litre volumetric flask

Procedure for Estimating Total Chloride Capacity

1. Put the resin into the column and ensure full conversion to the chloride form by regenerating with 500 ml of 0.5 N HCl at 1 BV/2 minutes. Rinse with 0.001 N HCl until 100 ml of the eluate requires not more than 1.5 ml of 0.1 N $AgNO_3$, using potassium chromate as indicator.
2. Transfer the resin to a measuring cylinder and tap it down under water. Measure and note the volume of resin. Carefully put the resin back into the column.
3. Pass 750 ml of 1% sodium nitrate solution through the resin at 1 BV/2 minutes and collect the eluate in a 1 litre volumetric flask. Rinse the resin with 200 ml deionized water. Dilute the flask contents to 1 litre with deionized water.
4. Titrate a suitable aliquot (*e.g.* 50 ml) with 0.1 N $AgNO_3$ using K_2CrO_4 indicator.

Let

$$\text{Resin volume} = V \text{ ml}$$

$$\text{Aliquot} = A \text{ ml}$$

$$\text{Normality AgNO}_3 = N$$

$$\text{Titre AgNO}_3 = T \text{ ml}$$

$$\text{Capacity} = \frac{TN}{AV} \times 1\,000\,000 \text{ meq/litre of chloride form resin}$$

TIME REQUIRED – 3 hours

Procedure for Estimating Capacity of Strong Groups
1. Use same column of resin as for total chloride capacity.
2. Convert the resin back to the chloride form using 500 ml of 0.5 N HCl, and rinse with deionized water at 1 BV/2 minutes.
3. Pass 100 ml of 2.5 N ammonium hydroxide solution through the resin, and rinse with deionized water until chloride and ammonia free.
4. Put 1 litre of 2.5% sodium chloride through the resin at 1 BV/minute. Rinse until chloride free.
   Note: The ammonia regenerates the weak base groups on the resin to the 'free base' form, but theoretically very few chloride ions should be removed from the strong base groups as ammonia is not a strong enough alkali for this purpose. However, in practice a few of the chloride ions on the strong base groups are removed and must be replaced by passing a neutral chloride solution through the resin. The weak base groups remain in the 'free base' form.
5. Continue as described in paragraphs 3 and 4 under 'Procedure for Estimating Total Chloride Capacity'.

The difference between the total chloride capacity and the capacity of strong base groups is the weak base capacity.

TIME REQUIRED – 2 hours 15 minutes

Salt Splitting Capacity

Salt splitting capacity is a value sometimes used to describe the strongly functional component only of a resin's total capacity. The term is usually reserved for strongly basic anion exchangers where it is identical to the value for strong base capacity or for weakly acidic cation exchange resins in which case it is a measure of the usually low capacity for the exchange of neutral cations.

Thermal Stability

The typical maximum operating temperatures allowed for resins are listed in Tables 4.2 and 4.3 which relate to almost all normal applications. At temperatures higher than the permissable maxima the decomposition characteristics depend upon the type of resin, its ionic form, and whether in a dry or aqueous environment.

Extreme pyrolysis, at around 600 °C, of dry styrenic sulfonic acid resins in the hydrogen form yields sulfur dioxide, various aromatic hydrocarbons, and carbonaceous residues. Strongly basic anion exchangers in the dry hydroxide form, on the other hand, decompose to give nitrogen oxides, hydrogen chloride, and hydrogen cyanide, besides various aromatic products. It is interesting that, in the calcium form, styrenesulfonic acid resins may be destructively oxidized to give pure calcium sulfate, and this can be made the basis of an accurate technique to determine dry weight capacity. At elevated pressures and temperatures in an aqueous environment cation exchange resins are hydrolysed to give acidic solutions, whereas anion exchangers give solutions of variable pH depending upon resin type and ionic form.

The severe conditions described above are abnormal in that for usual applications of ion exchange the most likely operational temperature span is 5 °C up to about 90 °C. Over such a temperature range the stability characteristics of resins in common use may be summarized as follows:

Cation Exchange Resins. Cation exchange resins are generally quite stable, especially if in a salt form.

Anion Exchange Resins. Strong base and weak base materials are also most stable in their salt forms compared with the hydroxide and free base forms respectively. The effect of increased temperature is to accelerate the loss of exchange capacity, in which respect acrylic anion exchangers whether weak or strong base are significantly more unstable that their styrenic counterparts. In the case of hydroxide form strong base anion exchange resins the predominant degradation route is either, or both, the loss of the functional group liberating the free amine (*e.g.* equation 4.22), or transformation from strong base to weak base with the formation of an alcohol as a reaction product (*Hoffman Degradation* – *e.g.* equation 4.23). The dominant degradation reactions for Types 1 and 2 strong base anion exchangers are illustrated by the following equations which also show the approximate percentage of total decomposition represented by each process:

Strong Base: Type 1

$$\underset{\text{strong base, Type 1}}{RCH_2N(CH_3)_3{}^+OH^-} \xrightarrow{50-80\%} \underset{\text{phenylmethanol}}{RCH_2OH} + \underset{\text{trimethylamine}}{N(CH_3)_3} \quad (4.22)$$

Properties and Characterization of Ion Exchange Resins

and

$$\xrightarrow{20-50\%} \underset{\text{weak base}}{\text{RCH}_2\text{N}(\text{CH}_3)_2} + \underset{\text{methanol}}{\text{CH}_3\text{OH}} \quad (4.23)$$

Strong Base: Type 2

$$\underset{\text{strong base, Type 2}}{\text{RCH}_2\text{N}(\text{CH}_3)_2(\text{C}_2\text{H}_4\text{OH})^+\text{OH}^-} \xrightarrow{80\%}$$

$$\underset{\text{weak base}}{\text{RCH}_2\text{N}(\text{CH}_3)_2} + \underset{\text{ethane-1,2-diol}}{\text{CH}_2\text{OH.CH}_2\text{OH}} \quad (4.24)$$

and

$$\rightarrow \underset{\text{weak base}}{\text{RCH}_2\text{N}(\text{CH}_3)(\text{C}_2\text{H}_4\text{OH})} + \underset{\text{methanol}}{\text{CH}_3\text{OH}} \quad (4.25)$$

and

$$\rightarrow \underset{\text{phenylmethanol}}{\text{RCH}_2\text{OH}} + \underset{\text{dimethylethanolamine}}{\text{N}(\text{CH}_3)_2(\text{C}_2\text{H}_4\text{OH})} \quad (4.26)$$

Thermal breakdown in the manner described is, in effect, an accelerated form of the normal capacity degradation behaviour occurring with use for strong base anion exchange resins. For Type 1 resins the principal mechanism is that of amine release and therefore a reduction in both total capacity and strong base capacity. For Type 2 anion exchange resins, however, the conversion of strong base groups to weak base is more prevalent resulting in a loss of strong base capacity whilst still maintaining a high total exchange capacity as the values in Table 4.5 illustrate.

It is important to differentiate between *thermal degradation* and *thermal shock* in that whereas the former is essentially a chemical breakdown, the latter arises from cycling a resin between extremes of temperature which imposes physical strains leading to bead fracture.

Table 4.5 *Total capacity, strong base capacity, and weak base capacity values for new and used strongly basic anion exchange resins*

WVC (eq l_R^{-1})	Styrenic strong base resin, (Cl form)			
	Type 1		Type 2	
	new	aged 6 yr	new	aged 6 yr
Total	60	52	60	48
Strong	60	51	60	27
Weak	—	1	—	21

(WVC ≡ wet volume capacity)

Macroporous resins are best able to resist thermal shock whereas gel structures are reported to demonstrate a better resistance to thermal degradation. The superior thermal stability of gel resins, which increases with decreasing crosslinking, is believed to be due to the increasing elasticity of the matrix being able to absorb, and therefore mechanically degrade, the thermal energy. It is not desirable to allow resins to encounter frost or freezing conditions since rapid thawing out gives rise to thermal shock. Freezing as such does not usually affect the internal aqueous gel phase, it being a concentrated electrolyte, but any restriction of expansion arising from the freezing of the interstitial water will result in resin beads becoming crushed.

PHYSICAL SPECIFICATION

Physical Appearance

The vast majority of modern resins are produced by the addition polymerization route and take the form of spherical beads. A few condensation polymer products remain available and these have an amorphous granular appearance.

Gel resins: usually shiny beads, clear to transmitted light.
Macroporous resins: usually dull beads of opaque or translucent appearance.
Colour: Without previous experience the identification of a resin by means of colour could be very misleading and is not advised.

Resin Particle Size

The particle size and size distribution of resins may be determined by a variety of techniques such as mechanical sieving, elutriation, sedimentation, photometric extinction, and microscopic examination. The commercial production of resins results in essentially a uniform Gaussian distribution of bead sizes from which various required gradings are achieved by mechanical sieving. The bead size range of conventional products is 300 μm to 1200 μm with a true mean value of approximately 700 μm. Particle size characterization by sieving of wet-swollen or dried beads on a weight or volume basis usually reports the *effective size* and a *uniformity coefficient* which are defined as follows, and in Figure 4.2:

Effective size = mesh size (μm) retaining 90% of the sieved sample

Properties and Characterization of Ion Exchange Resins 83

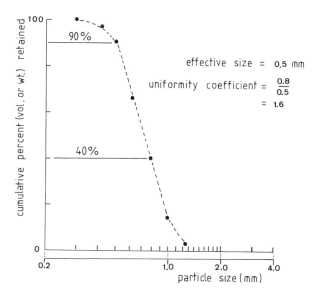

Figure 4.2 *Cumulative size distribution for wet sieved standard strong cation resin (H form)*

$$\text{Uniformity Coefficient} = \frac{\text{mesh size } (\mu m) \text{ retaining } 40\% \text{ of sample}}{\text{mesh size } (\mu m) \text{ retaining } 90\% \text{ of sample}}$$

Statistical methods may also be adopted to evaluate resin particle size distributions based upon population frequencies according to number (microscope count), specific surface area (photometric extinction methods), or weight–volume (sieving, sedimentation), providing the basis of the assessment is clearly stated. Since the rate of ion exchange is essentially related to the specific surface area of a resin, some workers have argued that the Harmonic mean particle size found from size distributions by weight or volume would be a preferable parameter since this places more emphasis on smaller resin beads. However sieving remains the most widely adopted technique for the production of standard particle size distributions of resins in common use.

Often the process design requirements for applications of ion exchange prescribe more rigorous particle size constraints which are met by either further sieving operations, or as is recently the case with some principal resin manufacturers, the ability to produce directly a uniformly sized product ($\pm 50\ \mu m$) at the resin copolymerization stage. The need for graded resins is required to meet the demands set by:

1. *Ion Exchange Chromatography*: This application can embrace a resin particle size from around 700 μm down to less than 10 μm.
2. *Deep Resin Beds*: A requirement for a narrow size distribution and sufficiently large effective size to minimize hydrodynamic pressure losses.
3. *High Specific Flowrates*: As for 2 above.
4. *Fast Kinetics*: A requirement for a narrow size distribution and small effective size in order to present a high specific surface area of resin.
5. *Designs Involving Resin Separation*: The separation of resins by elutriation benefits from there being a large density difference between the resins concerned, with the least dense component possessing a small effective size and narrow size distribution, whilst the heavier component could be expected to benefit from having a large effective size and narrow size distribution. In fact recent work identifies resin particle density as being the most important separability factor rather than large differences in particle size. Clearly, some requirements of a design considered together may be contradictory but as with most practical situations a compromise applies, which is usually well met by the broad range of resin gradings available from resin manufacturers.

Resin Density ($kg\,l^{-1}$)

The true density of any resin in its dry or swollen form will, of course, depend upon resin type, structure, degree of crosslinking, and ionic form. Precise values for the true density of swollen resins is an important consideration when predicting the hydrodynamic behaviour of continuous countercurrent systems or combinations of different resin types contained in fixed beds. The density of defined proprietary resin products is predictably fixed, but an increase in density may be achieved by the use of halogen substituted monomers during synthesis or copolymerizing a resin around a dense inert core; for example, magnetic resins which encapsulate a ferromagnetic material such as gamma iron(III) oxide, γ-Fe_2O_3.

Measurement of density for swollen exchangers is easily determined in water using pyknometric or simple density bottle methods (see Box 4.8); whilst a non-swelling agent, for example octane, may be used for

Table 4.6 Density values for some typical ion exchange resins in various ionic forms

Resin type	Matrix	Ionic form	Swollen density ($kg\,l^{-1}$) gel	macroporous
Strong Acid Cation	styrene–DVB	H^+	1.22	1.18
		Na^+	1.28	1.24
		NH_4^+	1.24	1.24
		Ca^{2+}	1.29	1.25
Weak Acid Cation	acrylic–DVB	H^+	1.19	1.18
		Ca^{2+}	1.20	1.16
Strong Base Anion	styrene–DVB	OH^-	1.08	1.04
		Cl^-	1.1	1.06
		SO_4^{2-}	1.13	1.12
	acrylic–DVB	Cl^-	1.07	1.08
Weak Base Anion	styrene–DVB	free base	1.1	1.04
	acrylic–DVB	free base	1.06	1.08

(values typical standard resins only)

determinations on dried resins. The densities of some typical cation and anion exchange resins are listed in Table 4.6 for selected ionic states, and deviation from an expected result provides useful evidence of possible structural degradation or resin fouling.

Bulk Density ($kg\,l^{-1}$)

This is commonly the value for the mass of a resin per unit apparent bulk volume as measured by recording the minimum settled (tapped or vibrated) volume of a weighed swollen resin in water.

BOX 4.8 Measurement of Swollen Resin Density

1. Apparent Resin Density

1. A measured 25 ml BV of resin is transferred to a 100 ml beaker and allowed to settle.
2. The excess water is decanted off, ensuring that no resin is lost.
3. The resin is dried in an oven overnight at $100 \pm 5\,°C$.
4. The resin is cooled in a desiccator.
5. The beaker plus resin are then accurately weighed (a g) and then the weight of the empty beaker is found (b g).

6. The apparent density $(D) = \dfrac{a-b}{BV}$ dry g/wet ml

Note: The value D may be used to derive or to check the total wet volume capacity (WVC). DWC meq/g × D g/ml = WVC meq/ml. Since the maximum error in WVC is the volume measurement, the determination of several D values to obtain a mean value will give a more accurate WVC determination than a direct measurement.

2. True Swollen Density

The true density or specific gravity of a swollen resin in a particular ionic form may be determined using a conventional density bottle method.

1. A dry density bottle plus stopper is weighed (a g).
2. Approximately 2–5 g of swollen resin is gently superficially dried on filter papers, transferred to the density bottle, and reweighed (b g).
3. The bottle is then filled with water of known ambient temperature, taking care to remove any entrapped air bubbles, and reweighed (c g).
4. Finally the weight of the bottle filled with water only is recorded (d g).

$$\text{Mass of resin} = (b-a) \text{ g}$$

$$\text{Mass of water equal in volume to the resin} = [(d-a)-(c-b)] \text{ g} = e \text{ g}$$

Therefore

$$\text{Specific Gravity} = \frac{b-a}{e}$$

and

$$\text{Density} = \left(\frac{b-a}{e}\right)\rho$$

where ρ is the density of pure water at the relevant temperature (consult tables).

The value for comparable swollen ionic forms will not be greatly different from the *shipping weight* which is the weight of swollen resin per unit apparent bulk volume measured after rinsing and subsequent draining.

A further useful resin density measurement is provided by the *apparent density* (Box 4.8) when defined as dry kilogram per litre minimum settled bulk volume (dry g per ml) since a check on capacity determinations is afforded by the relationship: DWC (eq/dry kg) × apparent density (dry $kg\,l^{-1}$) = WVC ($eq\,l^{-1}$). An unexpected change in the bulk density (dry $kg\,l^{-1}$) can also usefully signal any trend towards bead fouling, loss of capacity, or de-crosslinking. Typical approximate apparent density values for commercial styrenic strongly functional resins in their standard forms are 0.40 (dry $kg\,l^{-1}$) and 0.35 (dry $kg\,l^{-1}$) for cation and anion exchange resins respectively.

Percent Whole Beads

As obviously implied, this is a quality parameter reporting the proportion of unbroken beads.

Sphericity

Sometimes a resin manufacturer will list a value for 'sphericity' which is the percentage of perfectly spherical beads in a resin sample as determined by a technique based on allowing resin, dried to a free flowing condition, to roll down an inclined surface.

CAUTION: *Any spillage of resin beads (wet or dry) should be cleaned up immediately* since failing to do this and subsequent attempts to stand or walk on the affected area is very likely to result in a painful, but effective, proof of Newton's First and Third Laws of Motion!

SUMMARY

It seems pertinent that this chapter, above all others, should contain a summary since any reader new to the practice of ion exchange could be excused thinking that resin selection is a veritable minefield of uncertainty. Outside selection of the appropriate resin type (cation or anion) and its strength (weak or strong) there are no cast iron rules, and experience outweighs all other considerations. Given that the primary selection of a resin has been made, Table 4.7 summarizes guidelines concerning the options in relation to matrix, structure, and particle size grading.

One final point: hitherto the cost of ion exchange resins has received little comment, but obviously this is a major factor, all other

Table 4.7 *Idealized resin selection criteria*

Operational requirement	Selection criteria
Physical stability	premium gel, high crosslinking macroporous acrylic matrix
High flowrates; Low pressure drop; Deep beds	large beads narrow size distribution rigid matrix
Chemical stability	high crosslinking macroporous
High regeneration efficiency	low selectivity small beads low crosslinking weak function resins
High operating capacity	high wet volume capacity (WVC) low crosslinking small beads low selectivity
Resin separation	bead size density
Low 'leakage' (see Chapter 7)	a compromise between high selectivity and high regeneration efficiency
Fouling resistance (anion exchange resins)	macroporous acrylic matrix

factors being equal. A need to consider matrix modified resins (acrylic) and structure modified materials (macroporous) can increase resin cost anywhere between 16% and 150% compared with standard gel styrenic products depending upon whether comparing anion or cation resins respectively.

FURTHER READING

R. Kunin, 'Methods of Studying Ion Exchange Resins', in 'Ion Exchange Resins', Wiley, New York and London, 2nd Edition, 1958, Ch. 15, p. 320.

G. Kuhne, 'Standardisation of Test Methods for Ion Exchange Resins', in 'Ion Exchangers', ed. K. Dorfner, Walter de Gruyter, Berlin and New York, 1991, Ch. 1.2, p. 397.

J. G. Grantham, 'Ion Exchange Resin Testing', in 'Ion Exchange in Water Treatment', Duolite International Ltd, 1982, Section 14, p. 60.

American Society of Testing Materials, Volume 11.02, 1993.

R. Kunin, 'Ion Exchange Resins', Krieger, Melbourne, Florida, 1972.

Chapter 5

Ion Exchange Equilibria

INTRODUCTION

Since the pioneering studies of Thompson and Way, workers in the field of ion exchange have continued to study the equilibrium distribution of ions and solvent between an exchanger and the external solution. The results of such investigations naturally lead to enquiring why some species should be preferred by a given exchanger to others. Furthermore predicting such *affinity* or *selectivity* behaviour, even qualitatively, still remains one of the most fascinating and challenging aspects yet to be fully understood. The content of this Chapter is, for the sake of brevity, confined mainly to a discussion of organic resins and much of the complementary work carried out on inorganic ion exchange materials has to be omitted.

Both empirical and more fundamental relationships have been established to describe and predict equilibrium states for ion exchange systems. Virtually by definition, an empirical model found to suit a particular system being studied is unlikely to apply unchanged to all different circumstances since the empirical coefficients and exponents do not relate to fundamental properties of the ions, resin, and solvent concerned – otherwise the model would not be empirical.

It follows from this that an empirical approach relies heavily upon exhaustive experimentation. An example of an empirical approach often applicable to studies of ion exchange equilibria is a modified form of the Langmuir adsorption equation:

$$Y = \frac{aX}{bX + 1} \quad (5.1)$$

where Y and X are equivalent or mole fractions of a chosen counter-ion species in the resin and external solution respectively, whilst a and b are constants valid for a given system only. This discussion is not in any way meant to denigrate the worth of any empirical approaches to ion exchange equilibria, since not only have they contributed greatly to our understanding of the subject, but many real multi-component systems are so complex that a degree of empiricism or data fitting is sometimes necessary to formulate a usable model to scale up pilot engineering designs for working industrial plant. Also, an empirical component is often unavoidable in models describing the complex equilibrium and kinetic behaviour of industrial ion exchange processes.

Contrary to empirical approaches, a fundamental approach has value in that the results demonstrate the validity or otherwise of a particular mechanism or model chosen for the system. For example, the application of thermodynamics to an ion exchange system does not necessarily require the setting up of a physicochemical model, but eventually the results must still be interpreted in terms of the molecular forces acting within the system. Selected molecular models enable the mechanisms of ion exchange phenomena to be better interpreted, but their success must be measured in terms of predicted accuracy which in turn depends upon the validity of the model and the accessibility of the various molecular parameters. Ideally, the mathematical equations describing the perfect model would contain quantities which were derived from the known fundamental data for the components of the system.

Many of the molecular models which have extended the understanding of ion exchange phenomena have stemmed from the study of polyelectrolyte solutions and crosslinked gels. These interpret the free energy changes accompanying physicochemical reactions in terms of the 'electrostatic forces' and configurational entropy changes occurring in the system. Both the thermodynamic and molecular approaches have proved reasonably successful in their application to ion exchangers, but whatever approach is adopted, the present-day knowledge of the physical chemistry of electrolyte solutions falls short of that required to treat theoretically, or to understand perfectly, the highly concentrated gel–electrolyte constituting a typical ion exchange resin.

That the development of a perfect understanding of ion exchange phenomena at a molecular level should prove so challenging and elusive is not surprising when one considers the complex mechanistic steps involved, namely:

1. *Swelling*. An ion exchange resin will absorb polar solvents (commonly water) thereby swelling the matrix giving rise to a

configurational entropy change due to the stretching and realignment of the copolymer chains. At the same time as swelling occurs an internal osmotic pressure develops (swelling pressure) which, when at equilibrium, counteracts any further change in solvent uptake.

2. *Electrolyte Sorption.* In fact an ion exchanger does not totally exclude the sorption of co-ions from an external electrolyte solution, and therefore any ideal model has to account for all other permeant species over and above the exchange of counter-ions.

3. *Ion Exchange.* Finally, the ion exchange reaction itself changes the ionic state of a resin, not only from an ionic composition standpoint, but also in relation to the previously mentioned properties of swelling water, ion hydration, and co-ion uptake. Thus the simple act of ion exchange in reality involves energetic changes between initial and final states of a system which together must, when fully understood, explain the observed preference (*selectivity*) of one ion over another.

Thus whether such energetic changes are interpreted on a thermodynamic or molecular basis, an appreciation of ion exchange equilibria reduces to explaining a complex set of interactions between for example:

1. Counter-ion and fixed ion ('coulombic', ion pairing)
2. Counter-ion and matrix (London and van der Waals forces)
3. Ion–solvent (solvation, hydration)
4. Solvent–solvent (mixed solvent systems)
5. Ion–dipole (polarization effects)

Interdisciplinary contributions from studies of ion exchange in glasses, zeolites, polyelectrolytes, biological systems, as well as resins have all helped to further our understanding of ion selectivity and formulate predictive fundamental models. Controversy still persists as to the relative contributions made by different interactions, but a loose summary would be that coulombic interactions of one type or another and hydration energies are considered to be significant for the exchange of simple cations whilst for the exchange of anions, water structure promoted increase in the entropy of exchange is thought to be a principal selectivity-determining factor. For large organic ions resin matrix–ion interactions add yet a further complicating factor.

SWELLING PHENOMENA AND THE SORPTION OF SOLVENTS

The electrochemical potentials of freely permeant components, ions and solvent, in chemical equilibrium are equal in each phase. The total electrochemical potential or partial molar free energy of a component is made up of contributions from its activity a, pressure P, and its ionic, electrical potential ψ, all referred to the same standard and reference states of unit activity $(a = 1)$, and infinite dilution (activity coefficient $= 1$) respectively at 1 atmosphere pressure. The electrochemical potential η of a component i in a given phase is given by the equation:

$$\eta_i = \mu_i^0 + RT \ln a_i + (P - 1)\bar{V}_i \pm z_i F \psi_i \qquad (5.2)$$

where for the component i, μ_i^0 is its chemical potential at the standard state of unit activity and 1 atmosphere pressure, a its activity, $\bar{V}_i$ its partial molar volume, $\pm z_i$ its electrovalency, and F is the Faraday constant. The above equation is important as it is the starting point for considering *Gibbs–Donnan* equilibria in ion exchange systems.

Some further nomenclature is now necessary to describe absorption equilibria in ion exchange systems. For a species i, m_i and C_i represent the molal and molar concentrations respectively, whilst N_i and X_i denote the mole fraction and equivalent ionic fraction of i respectively. Single ion activity coefficients are denoted γ_i and mean ionic activity coefficients by $\gamma_i\pm$. Whether the latter quantities refer to the molar or molal concentration scales is decided by the choice of units defining concentration. Thermodynamic activities and activity coefficients for the resin phase using the equivalent or mole fraction concentration scale (rational scale) are sometimes defined differently and are discussed in a later section. Finally, the exchanger and external solution phases are differentiated by subscripts r and s respectively.

When resinous exchangers are placed in a solution containing an electrolyte, some or all of several features may be observed. Firstly, the resin exchanger swells by imbibing solvent from the external phase. Secondly, some electrolyte penetrates the exchanger to an extent which cannot be explained by the stoichiometry of the ion exchange reaction. Thirdly, when an exchanger phase and an external electrolyte solution coexist, an ion exchange reaction usually, but not necessarily, takes place. All three processes can occur at the same time and obviously influence the distribution of freely diffusible species in the system.

Swelling and Water Sorption

In 1948 an approach to ion exchange phenomena which has had a pronounced influence upon subsequent thinking was put forward by Gregor. The swelling properties of the resin and the equilibrium distribution of counter-ions were explained in terms of a Gibbs–Donnan membrane equilibrium and a mechanical model for the resin structure. The mechanical component of Gregor's approach is shown in Figure 5.1 where the crosslinking of the exchanger is regarded as perfectly elastic springs. The chemical components of the system are the *solvated* permeant ions, the *matrix* with its *solvated* fixed ions, and the *free solvent*.

The resinous exchanger may be regarded as a salt which becomes hydrated as the resin swells and imbibes water. The counter-ions dissociate, thus forming a concentrated electrolyte solution within the resin. The osmotic activity of the internal gel electrolyte causes further amounts of water to enter the resin phase which therefore continues to swell. This swelling of the resin is accompanied by a stretching of the crosslinked hydrocarbon matrix, or in terms of Gregor's model, an extension of the elastic springs. The net restoring force of the extended springs (crosslinks) is interpreted as an internal swelling pressure acting on the pore liquid within the resin. Equilibrium results when the osmotic forces counterbalance the mechanical restoring forces (*swelling pressure*) of the matrix.

Alternatively the molecular models for linear and crosslinked polyelectrolytes (including ion exchange resins) interpret the swelling

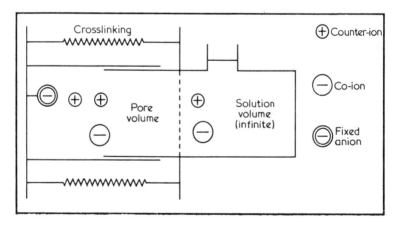

Figure 5.1 *Gregor's model of an ion exchange resin*
(Reprinted with permission from H. P. Gregor, *J. Am. Chem. Soc.*, 1951, **73**, 642. © 1951 American Chemical Society)

phenomenon, not in terms of a mechanical model, but rather as the mutual compensating effects of electrostatic interactions and changes in the configurational entropy of the charged polymer chains.

Gregor and Pepper and their co-workers have carried out detailed studies on the swelling properties of ion exchange resins from which the following general conclusions may be drawn:

(a) Swelling is more pronounced if the fixed ion and the counter-ions show a high tendency for solvation. It is difficult to generalize about the swollen volumes of ion exchangers in different ionic forms, but when values are compared on an equivalent basis some definite trends may be observed. For example, the equivalent swollen volumes (V_e) of polystyrenesulfonate resins increase as they are saturated with cations of increasing hydrated radius. Table 5.1 gives some experimental values for V_e, W_e, the specific dry volume V_m, and the resin phase molality m for various ions on a sulfonic acid resin. On the basis of ion size the specific volumes for the silver and thallous forms of the resin appear anomalous and indicate some type of ion association in the resin phase. Abnormal specific or equivalent resin volumes have been reported for some tetra-alkylammonium salt forms of polystyrenesulfonate exchangers, and also for strong base anion exchangers carrying large organic anions. These observations

Table 5.1 *Specific values of swollen volume (V_e), swollen weight (W_e), dry volume (V_m), and resin phase molality (m) for different ionic forms of a standard strongly acidic cation exchange resin (8% DVB)* (from H. P. Gregor, F. Gutoff, and J. Bregman, *J. Colloid Sci.*, 1951, **6**, 245)

Cation	$V_e(\text{cm}^3)$	$W_e(\text{g})$	$V_m(\text{cm}^3)$	m
H	1.524	1.880	0.696	5.6
Li	1.503	1.898	0.702	5.6
Na	1.418	1.870	0.730	6.4
K	1.357	1.843	0.763	7.4
Cs	1.391	2.292	—	7.8
Ag	1.209	2.094	0.685	10.7
Tl	1.182	2.492	—	10.3
Mg	1.456	1.921	0.707	5.7
Ca	1.372	1.853	0.720	6.4
Sr	1.345	1.938	—	6.8
Ba	1.264	1.946	0.726	7.9

clearly indicate that some type of specific interaction operates in these systems which are further discussed later when considering selectivity in ion exchange systems.

(b) The tendency of a resin to swell decreases with increasing crosslinking since the matrix becomes more rigid and the swelling pressure increases. This fact is of great importance when considering a resin for practical applications and industrial processes since the rate of ion exchange and its dynamic equilibrium characteristics are both adversely affected if a resin is too de-swollen to allow unimpaired reversible diffusion of ions.

(c) Swelling is favoured by an increase in dilution of the external solution since this increases the difference in the osmotic activity between the two phases. Similarly, swelling is favoured by an increase in the exchange capacity of the resin.

(d) In the absence of ion association effects, the number of mobile counter-ions in the resin phase is greater the lower their valency. Therefore the degree of swelling of a resin decreases with increasing valency of the counter-ion, which is in keeping with osmotic activity being a colligative property.

The thermodynamic relationship describing the osmotic equilibrium between a resin and pure liquid water may be derived as a result of reconsidering equation 5.2. Equilibrium, with pure water only, means that ion exchange need not be considered and therefore the electrical potential term $\pm z_i F \psi_i$ can be ignored, whereby the chemical potential μ_i of component i is given by:

$$\mu_i = \mu_i^0 + RT \ln a_i + (P - 1)\bar{V}_i \qquad (5.3)$$

The permeant component in this case is pure liquid water and remembering that at equilibrium the chemical potentials in both external and resin phases are equal, and that $(\mu_i^0)_s$ equals $(\mu_i^0)_r$ for the same standard and reference states for both the resin and external phases, it may be shown that:

$$RT \ln \frac{(a_w)_s}{(a_w)_r} = RT \ln \frac{1}{(a_w)_r} = \Pi(\bar{V}_w) \qquad (5.4)$$

where $(a_w)_r$ is the water activity in the resin, Π is the swelling pressure ($\Pi = P_r - P_s$), and $\bar{V}_w$ is the partial molar volume of water in the resin which is assumed to be equal to that of pure water.

Equation 5.4 is the basis of a more detailed and fundamental study of the swelling process achieved through the study of resin–water vapour sorption isotherms obtained isopiestically (*i.e.* at equal total pressures and equal resin water content). The isopiestic vapour pressure technique takes account of variable activity of the water in the resin (and therefore Π) by allowing the resin to come to equilibrium with water vapour at different partial vapour pressures P_w. It is assumed that two resins of the same structural type, but with different degrees of crosslinking, have the same water activity at the same equivalent water content. At equilibrium between resin and vapour phases the water activity in resins (1) and (2) are given by:

$$RT \ln (a_w)_1 = RT \ln \left(\frac{P_w}{P_0}\right)_1 - \Pi_1 \bar{V}_w \qquad (5.5)$$

$$RT \ln (a_w)_2 = RT \ln \left(\frac{P_w}{P_0}\right)_2 - \Pi_2 \bar{V}_w \qquad (5.6)$$

where P_0 is the saturation vapour pressure of water, P_w/P_0 the relative humidity, and Π_1, Π_2 the osmotic pressure (swelling pressure) of resins 1 and 2 respectively. At equal resin specific water contents (mole equiv.$^{-1}$) the activity of water in both resins are deemed equal, which assumes no pressure dependence of the partial molar volume of water, $\bar{V}_w$. Hence applying these conditions:

$$RT \ln \frac{(P_w)_1}{(P_w)_2} = (\Pi_1 - \Pi_2) \bar{V}_w \qquad (5.7)$$

Resin 1 is that whose swelling pressure is required whilst resin 2 is a reference resin of extremely low crosslinking such that Π_2 is virtually zero, whereby:

$$RT \ln \frac{(P_w)_1}{(P_w)_2} \cong \Pi_1 \bar{V}_w \qquad (5.8)$$

At saturation with resin 1 fully swollen, the relative humidity (P_w/P_0) is equal to unity whilst the reference resin at the same specific water content cannot be at saturation and exhibits a lower partial pressure or relative humidity. Hence the swelling pressure is found from the relationship:

$$RT \ln \frac{1}{\left(\frac{P_w}{P_0}\right)_2} = \Pi_1 \bar{V}_w \qquad (5.9)$$

Water vapour sorption isotherms for a 10% crosslinked sulfonic acid cation exchanger and a 0.4% crosslinked reference resin are shown in Figure 5.2 together with a calculation of the swelling pressure. The anomalous relative positions of the curves at low values of relative humidity P/P_0 may be explained in terms of van der Waals type intramolecular forces which weakly bind adjacent polymer chains in de-swollen exchangers of low crosslinking.

Osmotic pressures of many hundreds of atmospheres are typical for commonly used resins and increase with increasing crosslinking. If the swelling of a resin is restricted the forces developed can be sufficient to effect mechanical damage and it is therefore important to anticipate the volume changes that might occur when cycling resins between differing ionic states and varying external ionic strengths and, *where weakly functional resins are concerned*, differing external pH. Similarly, shrinkage occurs when the osmotic driving force between a resin and external solution is lowered as for example when a strongly functional exchanger (dissociated functional group) is contacted by a concentrated strong electrolyte (Figure 5.3). Or, in the case of weakly functional resins, when a change in external pH causes ion association of the functional group.

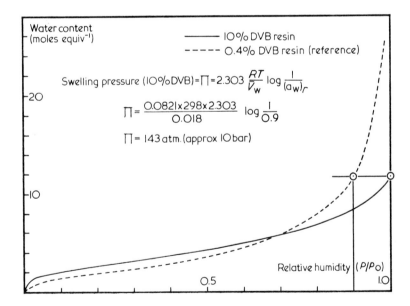

Figure 5.2 *Water vapour sorption isotherms and calculation of swelling pressure* (Reprinted with permission from B. R. Sundheim, M. H. Waxman, and H. P. Gregor, *J. Phys. Chem.*, 1953, **57**, 974. © 1953 American Chemical Society)

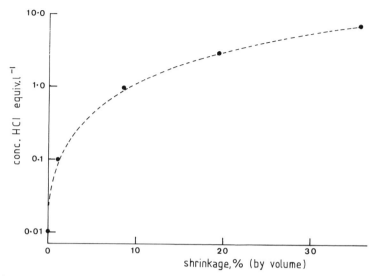

Figure 5.3 *Dependence of resin volume on the external electrolyte concentration for the hydrogen form of styrenesulfonate cation exchange resin (8% DVB) in hydrochloric acid solution*
(Data abstracted from W. C. Bauman and J. Eichorn, *J. Am. Chem. Soc.*, 1947, **69**, 2830)

Swelling pressure and resin hydration studies have been a significant landmark in the theoretical treatment of ion exchange resins. A linear relationship has been shown to exist between the equivalent volume of a strong acid or strong base exchanger and the swelling pressure. Such a relationship was independent of the ionic form and the equilibrium water activity in the external phase thus demonstrating that Gregor's mechanical model for the swelling behaviour is realistic for these types of resin.

Sorption isotherm determinations have been used to study the thermodynamics of resin hydration, and in ion exchange equilibrium studies for estimating resin phase activity coefficients from osmotic coefficient data. However, like activity coefficients, the complete determination of osmotic coefficients from fundamental data is not possible and predictive models employing osmotic coefficients are somewhat empirical, which further demonstrates our distance from fully understanding the behaviour of concentrated electrolyte solutions. Understanding the role played by solvent structure, especially water, in ion exchange resins has also been greatly assisted by spectral studies using infrared and nuclear magnetic resonance methods.

Sorption of Non-electrolytes

There is no general quantitative theory for the absorption of non-electrolytes by ion exchange resins. The osmotic activity of the resin phase arises from the dissociation of the counter-ions which is favoured by the uptake of polar solvents. In the absence of strong specific interactions between the resin and the solvent it has been observed that swelling is usually, but not always, greater in the more polar solvent, but it remains impossible to generalize since for some mixed solvents, for example ethanol–benzene (1:1) with a sulfonic acid cation exchange resin, the swelling from dry is greater than with either solvent alone. The same applies to absorption from mixed aqueous–organic solvent phases, where measured distribution coefficients generally show a preference for the more polar constituent.

Ideally, the distribution coefficient should be unity for the partition in mixed solvents, but the solvation of fixed ions and counter-ions in the resin phase exerts a salting out effect on the less polar solvent, thereby lowering its distribution coefficient. Distribution studies for many organic solvents between ion exchange resins and aqueous solutions indicate a coefficient of less than unity in most cases; an outstanding exception is phenol with strong base anion exchangers which is sorbed very strongly. The general trend of decreasing distribution coefficient with increasing external organic concentration and increasing crosslinking has been demonstrated for several aliphatic acids and alcohols in sulfonic acid resins. Under certain conditions the resin may exhibit an increasing preference for the organic component with increasing molecular weight along a homologous series.

Strong sorption with essentially non-polar solvents has been observed, and it is generally accepted that a salting in of the non-polar component can occur through London or van der Waals interactions between the organic solvent and the polymer structure of the resin, which brings about a decrease in enthalpy ΔH due to matrix–solvent interaction and decrease in entropy ΔS arising from the loss of rotational and translational degrees of freedom of the 'bound' solvent as described by Feitelson.

The enthalpy change is believed to dominate thereby giving a net reduction in the Gibbs free energy (ΔG) of the system in favour of sorption of the non-polar solvent. Conversely, water structure enhanced increase in system entropy could explain instances where preferential sorption of large organic species from aqueous–organic solvents are observed, but this topic is best left until discussing selectivity later in the chapter. Strong absorption effects are also to be expected if a non-polar species is able to complex with the counter-ions on the resin. For example, resins carrying transition metal counter-ions are able to sorb organic amines; and monosaccharides are absorbed by strong base anion exchange resins in the borate form.

SORPTION OF NON-EXCHANGE ELECTROLYTE AND THE DONNAN EQUILIBRIUM

When two coexisting phases are subjected to the condition that one or several of the ionic components cannot pass from one phase to the other, a particular type of equilibrium is set up, called a *Donnan equilibrium*. This type of ionic equilibrium is unique for ionic absorbents containing a fixed ion, and the isotherm describing such equilibria differs from the established Langmuir; Freundlich; and Brunauer, Emmett, and Teller types which are found for neutral species.

When, for example, a cation exchanger in the A^{z_A} form coexists with a solution of a strong electrolyte $A^{z_A}{}_{z_Y} \cdot Y^{z_Y}{}_{z_A}$, no net ion exchange can occur, but under certain conditions electrolyte from the external solution enters the resin phase (z_A = charge of counter-ion A). This phenomenon, and the extent to which it occurs, is explained by the sign and magnitude of the Donnan potential existing at the resin–solution interface. In the case of a cation exchanger initially containing no co-ions (Y^{z_Y}) being placed in the dilute electrolyte solution, the concentration differential across the resin–solution interface causes co-ions from the solution to diffuse into the resin, whilst counter-ions migrate into the solution from the exchanger. The fixed ion of the resin cannot migrate and a high electrical potential formed at the interface arrests any further diffusing tendency of the ions. It is important to realize that further co-ion uptake by the resin is accompanied by an equivalent amount of counter-ions to maintain macroscopic electrical neutrality. Thus electrolyte or co-ion uptake by the resin are synonymous terms.

For cation exchangers, the Donnan potential at the interface is negative (positive for anion exchangers) and the following important principles apply. The Donnan potential is greatest, and therefore co-ions more completely excluded, when the concentration difference between the resin phase and the external solution is large. Therefore a high exchange capacity, high crosslinking and increasing dilution of the external solution are factors that are favourable to the exclusion of electrolyte from the exchanger. Macroporous resins, because of their true porosity allow a greater uptake of non-exchange electrolyte compared with equivalent gel exchangers as shown in Figure 5.4.

Notice also that when comparing cation and anion exchange resins of the same matrix type, co-ion uptake is greatest for the anion exchanger because of their lower specific exchange capacity as predicted by Donnan Theory. The potential required to counteract the initial escape of the counter-ions is least for those ions of high valence, and the efficiency of electrolyte exclusion is correspondingly lower. Also a given Donnan potential excludes multivalent co-ions more efficiently than monovalent ions. An increase in the external electrolyte concentration favours the penetration of electrolyte into the exchanger as previously mentioned, but this is also accompanied by a deswelling (shrinking) of the resin due to a lowering of the water activity in the external solution relative to the exchanger phase.

Using equation 5.2 to equate the electrochemical potentials of the co-ion in both phases and equation 5.3, the following general equation

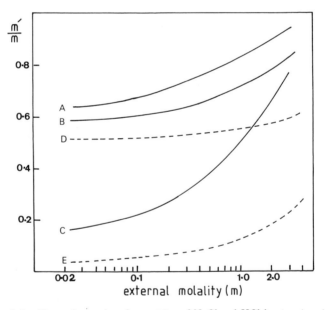

Figure 5.4 *Non-exchange electrolyte sorption of NaCl and HCl by styrenic anion and cation exchange resins respectively against external concentration*
[m' = internal molality; A = macroporous SBA (low capacity); B = macroporous SBA (conventional); C = isoporous SBA (gel); D = macroporous SAC; E = gel SAC]
(Data from J. R. Millar, 'Fundamentals of Ion Exchange', *Chem. Ind. (London)*, 1973, **5**, 409)

may be derived for describing Donnan co-ion equilibria in ion exchangers:

$$\left[\frac{(a_\pm)_r}{(a_\pm)_s}\right]^\nu = \left[\frac{(a_w)_r}{(a_w)_s}\right]^{\bar{V}_{AY}/\bar{V}_w} \tag{5.10}$$

or

$$\left[\frac{(m_Y)_r}{(m_Y)_s}\right]^{\nu_Y} = \left[\frac{z_Y(m_Y)_s}{z_Y(m_Y)_r + (m_R)_r}\right]^{\nu_A} \left[\frac{(\gamma_\pm)_s}{(\gamma_\pm)_r}\right]^\nu \left[\frac{(a_w)_r}{(a_w)_s}\right]^{\bar{V}_{AY}/\bar{V}_w} \tag{5.11}$$

where the dissociation of 1 mole of the electrolyte gives ν_A ions of A^{z_A} and ν_Y ions of Y^{z_Y}, and $\nu = \nu_A + \nu_Y$. The mean ionic activities and mean ionic activity coefficients of the electrolyte are denoted by $(a_\pm)$ and $(\gamma_\pm)$ respectively and its partial molar volume by $\bar{V}_{AY}$. The quantity $(m_R)_r$ is the molal concentration of fixed ionic groups within the resin. The influence of all the previously mentioned system variables on the co-ion uptake may be predicted from equation 5.11. For *dilute* solutions of simple 1:1 electrolytes where $(m_R)_r \gg (m_Y)_r$ and in the absence of any specific ion association effects in *either* phase, equation 5.11 may be written:

$$(m_Y)_r \cong (m_Y)_s^2 \left[\frac{z_Y}{(m_R)_r}\right] \left[\frac{(\gamma_\pm)_s}{(\gamma_\pm)_r}\right]^2 \left[\frac{(a_w)_r}{(a_w)_s}\right]^{\bar{V}_{AY}/\bar{V}_w} \tag{5.12}$$

If swelling effects are ignored, the usual simplified Donnan equation for co-ion uptake from dilute solutions is obtained namely:

$$(m_Y)_r \cong (m_Y)_s^2 \left[\frac{z_Y}{(m_R)_r}\right] \left[\frac{(\gamma_\pm)_s}{(\gamma_\pm)_r}\right]^2 \tag{5.13}$$

With the reference states defined earlier for the resin and solution phases, the activity coefficient ratio in the preceding equations should approach a constant value with increasing dilution of the external solution, and ultimately become unity with increasing dilution of both the resin and external phases. Intuitively, the latter condition would not be expected to hold since the reference state of infinite dilution for the exchanger is not compatible with the physical existence of a crosslinked ion exchanger. The observed result, which has been confirmed by a large number of researchers, is that the mean ionic activity coefficient of the absorbed electrolyte in the resin phase decreases with increasing dilution of the external electrolyte.

In other words, the degree of electrolyte uptake with increasing dilution of the external solution is greater than that which would be expected from the ideal Donnan equation. An activity coefficient ratio of less than unity in equation 5.13 is to be expected, but the experimental finding that this ratio increases with increasing dilution of the external solution is unexpected and the fundamental mechanism of this phenomenon is still the subject of great interest and debate. The results of earlier studies were interpreted in terms of specific ionic interactions in the resin phase, and Gregor refined his earlier model in order to account for the observed activity coefficient behaviour.

It was recognized that analytical errors, resin impurities, and heterogeneity in the structure of the resin might be contributory factors, and later experiments were designed to take account of these effects. Heterogeneity in the resin structure may be responsible, at least in part, for the experimentally measured power of the $(m_Y)_s$ term in equation 5.13 being found to be less than two. Interactions between co-ions in the resin phase, and variations in the electrical potential (ψ) of the absorbed co-ion because of inefficient screening of the fixed ions by the counter-ions, may be the cause of the mean ionic activity coefficient of the absorbed electrolyte decreasing with increasing dilution of the external solution. Obviously, constant values for the electrical potentials of ions are implicit in deriving equation 5.13. It must be concluded that the Donnan theory is valid only for homogeneous exchangers and constant electrical potentials within the resin, and under certain practical conditions it is not strictly applicable. Much progress has been made in unifying Donnan theory and observed ion diffusion behaviour through the work of Glueckauf, Schlögl and Schodel, and Mackie and Meares. Counter-ion association in the resin, specific interactions between the co-ions and the fixed ionic group, and complex ion formation in the external solution can all greatly influence the extent of co-ion sorption. In the case of mixed electrolyte–non-electrolyte solutions, the Donnan potential excludes the electrolyte preferentially and this is the basis of the separation of such mixtures by the technique of ion exclusion.

RELATIVE AFFINITY

Even before recent advances in the theoretical and mechanistic understanding of ion exchange equilibria it was appreciated that cations and anions demonstrated an order of preferred affinity towards uptake by conventional resinous exchangers. The following sequences represent the order usually found for dilute solutions of commonly encountered ions with standard resins.

Strong Acid Cation Resin (styrenic – sulfonate):

$$Ag^+ > Cs^+ > K^+ > NH_4^+ > Na^+ > H^+ > Li^+;$$
$$Ba^{2+} > Pb^{2+} > Ag^+ > Sr^{2+} > Ca^{2+} > Ni^{2+} > Cd^{2+} > Cu^{2+} >$$
$$Co^{2+} > Zn^{2+} > Mg^{2+} > Cs^+$$

Weak Acid Cation (acrylic – carboxylate):

$$H^+ \gg Cu^{2+} > Pb^{2+} > Ni^{2+} > Co^{2+} > Fe^{3+} > Ca^{2+} > Mg^{2+} >$$
$$Na^+ > K^+ > Cs^+$$

Strong Base Anion – Type 1 (styrenic – quaternary ammonium):

$$SO_4^{2-} > HSO_4^- > I^- > NO_3^- > Br^- > Cl^- > HCO_3^- >$$
$$HSiO_3^- > F^- > OH^-;$$
$$SO_4^{2-} > ClO_4^- > ClO^- > NO_3^-$$

Strong Base Anion – Type 2 (styrenic – quaternary ammonium):

The slightly lower base strength of Type 2 strong base anion exchangers results in the relative affinity of hydroxide ion usually coming between that of fluoride ion and hydrogencarbonate ion.

Weak Base Anion (styrenic – amine):

$$OH^- \gg SO_4^{2-} > HSO_4^- > I^- > NO_3^- > Br^- > Cl^- > F^-$$

Generally: for all cation and anion exchange affinities on resins,

$$ION^{|z+1|} > ION^{|z|}$$

where z is the electrovalency of the ion ($\pm$).

Note: See Box 4.3 concerning the conversion of anion exchange resins to the hydrogencarbonate and sulfate forms.

SELECTIVITY COEFFICIENT

The stoichiometric ion exchange reaction between two counter-ions A^{z_A} and B^{z_B} may be written:

$$z_B R_{z_A}.A^{z_A} + z_A B^{z_B} \rightleftharpoons z_A R_{z_B}.B^{z_B} + z_B A^{z_A} \qquad (5.14)$$

where z is the ionic charge and R is the resin fixed ion. This equation is a general one and applies to both cation and anion exchange; it represents the distribution of counter-ions A and B between the exchanger and external solution. In the early days of the subject, empirical and semi-empirical relationships derived from absorption or mass action type models were found to fit most exchange data. For the most part the early theoretical treatments have given way to the more formal treatments based on the law of mass action, thermodynamics, and molecular models.

Numerous investigators have shown that for an ion exchange reaction at equilibrium, the distribution of counter-ions between the two phases is not equal. Thus for a particular exchanger one ion is generally preferred over the other and the exchanger is said to exhibit *selectivity*.

The experimentally observed selectivity shown by an exchanger is often represented by the value of the *separation factor*, α_A^B, which is defined by the equation:

$$\alpha_A^B = \frac{(m_B)_r (m_A)_s}{(m_A)_r (m_B)_s} = \frac{(C_B)_r (C_A)_s}{(C_A)_r (C_B)_s} = \frac{(X_B)_r (X_A)_s}{(X_A)_r (X_B)_s} \quad (5.15)$$

A value of α_A^B greater than unity means that ion B is preferred by the exchanger for a given point on the exchange isotherm. Typical isotherm behaviour is shown in Figure 5.5 where curves 1 and 2 represent favourable ($\alpha_A^B > 1$) and unfavourable equilibria ($\alpha_A^B < 1$) respectively. The sigmoid isotherm, curve 3, represents quite common behaviour where the direction of preferred uptake by the exchanger changes as a function of resin loading for, in particular, high total external concentrations.

For theoretical treatments of ion exchange it is preferred to define equilibrium in terms of the *selectivity coefficient* (K_A^B), which is the mass action relationship written for the reaction according to a defined choice of concentration units:

$$K_{m\,A}^{\,B} = \frac{(m_B)_r^{z_A} (m_A)_s^{z_B}}{(m_A)_r^{z_B} (m_B)_s^{z_A}} \quad (5.16)$$

$$K_{c\,A}^{\,B} = \frac{(C_B)_r^{z_A} (C_A)_s^{z_B}}{(C_A)_r^{z_B} (C_B)_s^{z_A}} \quad (5.17)$$

$$K_{x\,A}^{\,B} = \frac{(X_B)_r^{z_A} (X_A)_s^{z_B}}{(X_A)_r^{z_B} (X_B)_s^{z_A}} \quad (5.18)$$

The above coefficients on different concentration scales are termed the

Ion Exchange Equilibria

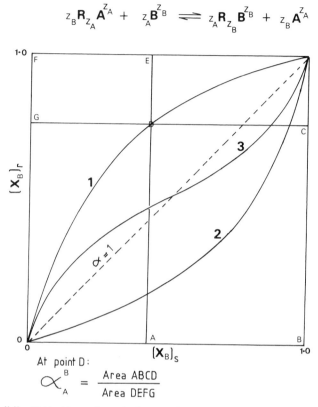

Figure 5.5 *Typical ion exchange isotherm profiles*

molal (m), molar (C), and rational (equivalent ionic fraction X) selectivity coefficients respectively. For the exchange of ions of equal valency all three coefficients have the same numerical value, and are related to the separation factor by the expression:

$$K_{\underset{A}{m}}^B = K_{\underset{A}{c}}^B = K_{\underset{A}{x}}^B = (\alpha_A^B)^{z_A} \tag{5.19}$$

In the case of heterovalent exchange the value of the selectivity coefficient depends upon the choice of the concentration units, and is related to the separation factor according to the following equation:

$$(\alpha_A^B)^{z_B} = K_{\underset{A}{m}}^B\left(\frac{(m_B)_r}{(m_B)_s}\right)^{z_B-z_A} = K_{\underset{A}{c}}^B\left(\frac{(C_B)_r}{(C_B)_s}\right)^{z_B-z_A} = K_{\underset{A}{x}}^B\left(\frac{(X_B)_r}{(X_B)_s}\right)^{z_B-z_A} \tag{5.20}$$

where $z_B > z_A$, and normally the concentration of ions B in the resin is greater than their concentration in the external solution. It is important to distinguish between the separation factor and selectivity coefficient for heterovalent ion exchange. Under ideal conditions and no inherent selectivity in the system, the selectivity coefficient would equal unity, yet from equation 5.20, α_A^B would always be greater than unity. This phenomenon has been termed 'electroselectivity' by Helfferich whereby the ion of the highest charge is preferred by the exchanger and this becomes more pronounced with increasing dilution of the external solution, and increasing $z_B - z_A$ for simple cations and anions.

The relation between the molar selectivity coefficient and the equivalent ionic fractions of ions A and B in both phases may be derived as follows:

$$K_A^B \left[\frac{(C_T)_r}{(C_T)_s} \right]^{z_B - z_A} = \frac{(X_B)_r^{z_A}(X_A)_s^{z_B}}{(X_A)_r^{z_B}(X_B)_s^{z_A}} = \frac{(X_B)_r^{z_A}[1 - (X_B)_s]^{z_B}}{[1 - (X_B)_r]^{z_B}(X_B)_s^{z_A}} \quad (5.21)$$

where $(C_T)_r$ and $(C_T)_s$ are the total equivalent concentrations of counter-ions in the resin and solution respectively. The phenomenon of electroselectivity has great significance in the practical application of ion exchange in that it accounts for the increase in magnitude of the separation factor for divalent over monovalent ions with decreasing total external concentration. This is especially relevant to water softening and demineralizing processes by ion exchange. A physical explanation for this effect lies in the action of the Donnan potential. For a given Donnan potential (negative) in a cation exchanger, multivalent cations would be attracted more strongly than monovalent cations, and the magnitude increases with dilution of the external solution.

BOX 5.1 A Simplified Experiment to Demonstrate Affinity Sequences for Ion Exchange Reactions and Estimation of the Selectivity Coefficient

1. Affinity Sequence

This experiment uses electrical conductivity as an approximate, but convenient, indicator of the attainment of equilibrium and degree of resin conversion for exchange involving the hydrogen ion. Of all cations, the hydrogen ion is the most highly conducting because of the unique charge transfer mechanism associated with hydrogen bonded hydroxonium ions H_3O^+. To a lesser extent the same property applies to hydroxide ions compared with all other anions. Thus consider the ion exchange reaction:

$$RH^+ + A^+ \rightleftharpoons RA^+ + H^+$$

Ion Exchange Equilibria

For a given *dilute* total external concentration, C_s (meq l^{-1}), and equilibration between constant initial quantities of resin (RH) and solution (AY) the net increase in solution conductivity will be greater, the greater the extent of exchange for ion A, *i.e.*, the greater the affinity of the resin for ion A.

Requirements

gel SAC resin (H form) swollen ~ 8–10% DVB
500 cm^3 volumes of 0.001 N solutions of HCl, LiCl, NaCl, KCl, CsCl, and CaCl$_2$, (C_s = 1 meq l^{-1})
beakers, 400 cm^3
conductivity meter
stirring rods (or magnetic stirring bench)
measuring cylinder (250 cm^3)

Procedure

Record the conductivity μ_{ACl} for each of the neutral salt solutions (250 cm^3) contained in separate beakers. Also record the conductivity of the 0.001 N HCl solution μ_{HCl}. Add exactly equal masses m g (about 0.2 g) of swollen (H form) resin to each solution, allow to reach equilibrium overnight with stirring, and note the final conductivity (μ_f).

Results

Assuming a linear relationship between conductivity and concentration for dilute solutions the equivalent ionic fraction of hydrogen ions in the external solution at equilibrium $(X_H)_s$ is given by:

$$(X_H)_s = \frac{(\mu_f - \mu_{ACl})}{(\mu_{HCl} - \mu_{ACl}) C_s}$$

List values of $(X_H)_s$ in order of decreasing magnitude for the exchange reactions tested which follows the affinity sequence:

$$Ca > Cs > K > Na > Li$$

2. Selectivity Coefficient

To *estimate* an apparent selectivity coefficient K_{app} the total exchange capacity of the H form resin in mequiv. per swollen gram is required. This value may be obtained by titration of the swollen resin or from the dry weight capacity DWC (Box 4.5) and the resin water content $W\%$ (Box 4.4), where:

$$\text{Swollen Weight Capacity } Q = \text{DWC} \left(1 - \frac{W\%}{100}\right) \text{meq g}^{-1}$$

Monovalent–Monovalent

$$RH^+ + A^+ \rightleftharpoons RA^+ + H^+$$

$$(K^{A^+}_{H^+})_{app} = K^{A^+}_{c\ H^+} = \frac{V(X_H)_s C_s/mQ}{[1-(V(X_H)_s C_s/mQ)]} \cdot \frac{(X_H)_s}{(1-(X_H)_s)}$$

where
m = mass of swollen resin (g)
V = volume of solution (l)
Q = swollen weight capacity (meq g^{-1})
$(X_H)_s$ = equiv. ionic fraction (H$^+$) in solution at equilibrium
C_s = total external concentration (meq l^{-1})
C_r = total resin phase concentration (meq l^{-1})

Monovalent–Divalent (See Text)

$$2RH^+ + A^{2+} \rightleftharpoons R_2A^{2+} + 2H^+$$

$$(K^{A^{2+}}_{H^+})_{app} = K^{A^{2+}}_{c\ H^+} \cdot \left(\frac{C_r}{C_s}\right) = \frac{V(X_H)_s C_s/mQ}{[1-(V(X_H)_s C_s/mQ)]^2} \cdot \frac{(X_H)_s^2}{(1-(X_H)_s)}$$

Note: For accurate work exact analysis of both the resin and external phases is undertaken, and account taken of co-ion uptake and changes in resin water content. For this approximate experiment, the actual solution phase composition at equilibrium may be determined by separately preparing a calibration of conductivity against (H – A) composition or by analysis for the ions concerned.

TIME REQUIRED – overnight

Selectivity coefficients are not generally constant over the whole exchange isotherm since their definition incorporates concentrations rather than activities. The relation between the thermodynamic exchange constant (K_{Th}) and the mass action constant on a particular concentration scale is obtained by introducing activity coefficients into the expression for the selectivity coefficient, thus:

$$K_{Th} = \frac{(m_B)_r^{z_A}(m_A)_s^{z_B}}{(m_A)_r^{z_B}(m_B)_s^{z_A}} \cdot \frac{(\gamma_B)_r^{z_A}(\gamma_A)_s^{z_B}}{(\gamma_A)_r^{z_B}(\gamma_B)_s^{z_A}} \qquad (5.22)$$

The value of the activity coefficient ratio in the external solution may be obtained from tabulated data for mean ionic activity coefficients in solution, or from standard electrolyte theory. Thus equation 5.22 may be corrected for the solution phase activities to read as follows:

$$K_{Th} = (K_A^B)_{corr} \left(\frac{\gamma_B^{z_A}}{\gamma_A^{z_B}}\right)_r \qquad (5.23)$$

The activity coefficient ratio in the resin phase which is the important selectivity determining factor is not obtainable by conventional means, but its determination by indirect methods forms the basis of understanding and predicting selectivity by the conventional thermodynamic approach.

When the standard and reference states for the exchanger and external solution phases are defined according to the conventional theory of electrolyte solutions, the thermodynamic exchange constant is by definition equal to unity. Therefore from equation 5.23 the observed selectivity in dilute solutions arises from the activity coefficient ratio in the exchanger phase thus:

$$K_{m\ A}^{\ B} = \left(\frac{\gamma_A^{z_B}}{\gamma_B^{z_A}}\right)_r \qquad (5.24)$$

In other words the complex behaviour of various ion–ion, and ion–solvent interactions are reflected in the abnormal values of the activity coefficients.

RATIONAL THERMODYNAMIC SELECTIVITY

If the standard and reference states for the exchanger phase are defined differently, whilst maintaining the conventional states for the solution phase, a thermodynamic selectivity scale can be set up for various ions where the value of the exchange constant indicates the degree of selectivity, as first demonstrated by Bonner, Argensinger, Hogfeldt, and others.

In this treatment the components of the exchanger phase are the mixed swollen resinates or compounds $R_{z_A}A^{z_A}$ and $R_{z_B}B^{z_B}$ whilst the standard and reference states are the pure salt forms of the resin in equilibrium with water. Thus the resin is regarded as a solid solution, but ideal behaviour is not assumed, and activity coefficients f are introduced to maintain thermodynamic rigour. If equivalent ionic fractions are used to express the composition of the resin phase, and the molal scale of concentration is retained for the solution phase, the *rational* thermodynamic equilibrium constant ($^N K_{Th}$) is given by:

$$^N K_{Th} = K' \left(\frac{f_{RB}^{z_A}}{f_{RA}^{z_B}} \right)_r \quad (5.25)$$

where K' is the corrected selectivity coefficient, and the activity coefficients f_{RA} and f_{RB} equal unity for the pure A and B forms of the resin respectively. Application of the Gibbs–Duhem equation to the resin phase plus further algebraic manipulation give the following expressions for the exchange constant and the individual activity coefficients:

$$\ln {}^N K_{Th} = (z_A - z_B) + \int_{X_B=0}^{X_B=1} \ln K' \, dX_B \quad (5.26)$$

$$\ln f_{RA}^{z_B} = -(z_A - z_B) X_B + X_B \ln K' - \int_{X_B=0}^{X_B} \ln K' \, dX_B \quad (5.27)$$

$$\ln f_{RB}^{z_A} = (z_A - z_B) X_A - X_A \ln K' + \int_{X_B}^{X_B=1} \ln K' \, dX_B \quad (5.28)$$

The abstract thermodynamic treatment outlined above resembles Kielland's approach to ion exchange equilibria on aluminosilicates, but unlike the latter case no simplifying assumptions are made concerning the relationship between the concentrations of the resin phase components and their activity coefficients.

Graphical integration methods are used to evaluate equations 5.26 and 5.27 which thus establish a scale for selectivity in ion exchangers. A typical selectivity series for some common cations on a sulfonic acid resin exchanging against Li^+ is shown in Table 5.2.

Multicomponent Systems

The vast majority of ion exchange equilibrium data has been obtained for binary ion systems. However many process applications deal with multicomponent systems, for example, water treatment, hydrometallurgical processing, and chromatographic separations. To be able to predict various equilibrium conditions for these more complex situations is of immense value to chemists and chemical engineers involved in process performance calculations to optimize plant designs. A rigorous theoretical approach to predicting multicomponent equilibria remains tedious as does establishing hard experimental data for every system that might apply.

Table 5.2 *Rational thermodynamic selectivity constant $^N K_A^B$ for cation B against Li^+ ion A on variously crosslinked polystyrenesulfonate cation exchange resins* (from O. D. Bonner and L. L. Smith, *J. Phys. Chem.*, 1957, **61**, 326)

Counter-ion B	Degree of Crosslinking (% DVB)		
	4%	8%	16%
H	1.32	1.27	1.47
Na	1.58	1.98	2.37
K	2.27	2.90	4.50
Cs	2.67	3.25	4.66
Ag	4.73	8.51	22.9
Ca	4.15	5.16	7.27
Ba	7.47	11.5	20.8
Cu	3.29	3.85	4.46

Streat and his co-workers have presented successful graphical methods whereby predictive triangular equilibrium diagrams for ternary cation systems may be derived from widely published binary data. An example of such a plot is shown in Figure 5.6 where the grid intersections give the predicted equilibrium position whilst the points represent actual measured values. The resin phase equivalent ionic fractions y are found from binary equilibrium data and plotted on a triangular grid for various calculated constant values of the solution phase composition x for a given component, giving a series of contour lines whose intersection gives the ternary equilibrium composition of the resin. An alternative graphical treatment may be adopted for mixed valency systems. Overall agreement between predicted and actual results is good thereby providing a useful technique for generating usable equilibrium data for process design calculations.

PREDICTION AND INTERPRETATION OF SELECTIVITY

The prediction of selectivity for a given ion over another even in qualitative terms, let alone quantitatively, ultimately requires an understanding of the ion exchange phenomenon in terms of the fundamental properties of the system components. No single characteristic can account for observed results, and studies to date amply demonstrate that many system properties affect selectivity behaviour in ways that have assisted our understanding of the mechanism involved.

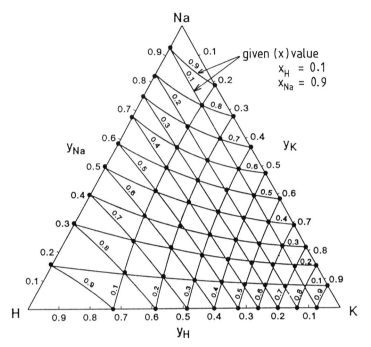

Figure 5.6 *A representation of ternary cation exchange equilibria, K^+–Na^+–H^+ system, where y = resin phase composition, and x = solution phase composition*
(Reproduced by permission from M. J. Slater, 'Principles of Ion Exchange Technology', Butterworth–Heinemann, Oxford, 1991)

Thermodynamic Approach

One of the earliest, and reasonably successful, approaches to quantitatively predicting selectivity behaviour was through the thermodynamic treatment of ion exchange systems as a *Gibbs–Donnan* membrane equilibrium. Such a description is given by equation 5.29 which for the sake of simplicity is shown in terms of single ion activity coefficients:

$$\ln\left[K_m^B{}_A\left(\frac{\gamma_B^{z_A}}{\gamma_A^{z_B}}\right)_r\left(\frac{\gamma_A^{z_B}}{\gamma_B^{z_A}}\right)_s\right] = \frac{\Pi}{RT}[z_B\bar{V}_A - z_A\bar{V}_B] \quad (5.29)$$

A rigorous thermodynamic expression would be required to consider terms for the equilibrium transport of water (solvent) and non-exchange electrolyte but these contributions are often ignored for exchange at low total external ionic strength.

All subsequent models for selectivity behaviour are, in some way or other, disguised in equation 5.29; the reason for this being that deviations from theoretical ideal behaviour as expressed through the values of resin phase activity coefficients, or by the 'interaction' energetics required by a mechanistic model, are equivalent statements. In other words, the fundamental causal factors which determine selectivity plus any inadequacies in our understanding are all reflected in the adopted model whether thermodynamic or molecular.

If the conventional standard and reference states for dilute electrolyte solutions are adopted the pressure–volume term is usually neglected for simple non-hydrated ions, and further assuming the activity coefficient ratio in the external solution to equal unity, equation 5.29 becomes:

$$K_{m\ A}^{\ B} = \left(\frac{\gamma_A^{z_B}}{\gamma_B^{z_A}}\right)_r \tag{5.30}$$

This simple relationship was derived before as equation 5.24, and was first used by Bauman and Eichorn in 1947 to predict selectivity sequences for simple monovalent cations from mean ionic activity coefficient data for pure aqueous electrolyte solutions containing a common anion. The inaccessibility of resin phase activity coefficients to direct measurement always remains a problem with thermodynamic equilibrium treatments. Therefore Glueckauf and others developed weight swelling and isopiestic water vapour sorption techniques to determine osmotic coefficients of pure salt forms of a resin, from which the mean ionic activity coefficients of mixed 'resinates' could be computed using a modified form of Harned's Rule. Such studies predicted selectivity coefficient values which were in fair agreement with experiment and also demonstrated the fixed ion of the resin to be osmotically inactive.

An even more thermodynamically rigorous approach was undertaken by Myers and Boyd in 1961 which avoided the use of empirical relationships such as Harned's Rule and gave predicted selectivity coefficients for alkali metal cation exchange on styrene sulfonic acid exchangers in fair agreement with observed values, particularly for resins of low crosslinking. An extension of these studies revealed that whilst the selectivity coefficients for pairs of simple monovalent cations expectedly approach unity in resins of low crosslinking, for the halogen anions on strong base anion exchange resins significant affinity differences remained. This result has consequently been shown to be very significant in that it predicts that the underlying 'interactions' governing resin affinities could be different for cations and anions.

During the period 1950–1970 the thermodynamic equilibrium treatments of ion exchange equilibria advanced at a rapid rate incorporating ever greater theoretical 'exactness', but in achieving this, whilst such progress became a very valuable theoretical contribution, it demanded an experimental burden that rather diminished its ease of application for casual predictive purposes. Over the same period other workers were formulating molecular models to describe the observed ion affinity sequences in terms of the energetics of the exchange path.

Before summarizing current opinion arising from this approach it is important to realize that classical thermodynamics does not require a model and is not concerned with the exchange *path* of an ion exchange reaction, but rather the energy *change* between initial and final states of a system. This is achieved through the measurement of the extensive properties of Gibbs free energy ΔG, enthalpy ΔH, and entropy ΔS. Whilst some controversy exists as to the exact interpretation of the entropy change, undoubtedly the study of equilibria through the rational thermodynamic equilibrium constant and calorimetrically determined enthalpy and heat capacity ΔC_p changes have contributed greatly to complementing, and discriminating, between several proposed mechanistic molecular models in their ability to account for the observed standard enthalpy $\Delta H^\ominus$ and entropy $\Delta S^\ominus$ changes during ion exchange reactions.

Energetics of Ion Exchange

The standard enthalpy change accompanying ion exchange reactions in resins is usually quite small, typically about $\pm (0.1-0.5 \text{ kJ eq}^{-1})$, and usually exhibits a minor dependence on temperature as given by the familiar van't Hoff equation:

$$\frac{d \ln {}^N K_{Th}}{d\left(\frac{1}{T}\right)} = -\frac{\Delta H^\ominus}{R} \qquad (5.31)$$

where T is the absolute temperature (K), R the molar gas constant (J mole^{-1} K^{-1}), and ${}^N K_{Th}$ the rational thermodynamic equilibrium constant. The measurement of the enthalpy changes accompanying ion exchange reactions may, in principle, be obtained from the temperature dependence of the equilibrium constant and a knowledge of the temperature dependence of activity coefficients. Fortunately it is easier to measure the standard enthalpy and heat capacity changes by direct calorimetry. Partial heat changes Q for various degrees of ion substitution gives the differential enthalpies of exchange,

$$\Delta \bar{H} = \frac{dQ}{dX},$$

as shown in Figure 5.7.

The differential enthalpy profile is usually seen to be a smooth function of resin loading which further supports the view that the

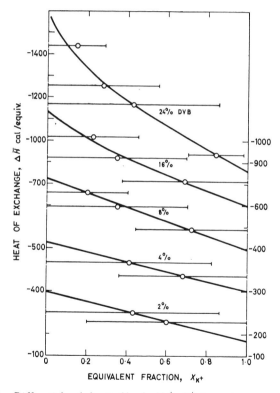

Figure 5.7 *Differential enthalpy profiles for Na^+-K^+ ion exchange on styrenesulfonate cation exchange resins*
(Reproduced by permission from G. E Boyd, 'Thermal Effects In Ion-Exchange Reactions With Organic Exchangers: Enthalpy and Heat Capacity Changes', in 'Ion Exchange In The Process Industries', Society of Chemical Industry, London, 1970, p. 261)

resin phase is a continuous, yet heterogeneous structure. This is unlike the situation often found for some zeolites where regions of distinctly different crystallinity give rise to discontinuities in the differential heats of exchange (Figure 5.8). Integration of the differential enthalpy plot against equivalent ionic fraction loading and correction for heats of dilution in the aqueous phase gives the standard enthalpy change $\Delta H^\ominus$. The standard free energy change $\Delta G^\ominus$ may be derived from the rational equilibrium constant data, whereby the standard entropy change $\Delta S^\ominus$ is found from the relationship:

$$\Delta G^\ominus = \Delta H^\ominus - T\Delta S^\ominus \qquad (5.32)$$

Several workers have reported thermochemical data for ion exchange in resins but a relatively recent extensive study has been given by Boyd, some of whose results are presented in Tables 5.3a–d. These and similar studies offer a means of attempting to unify the thermodynamics of ion exchange with the exchange mechanism as postulated by various molecular theories.

Clearly, any mechanistic theory or molecular model has to account

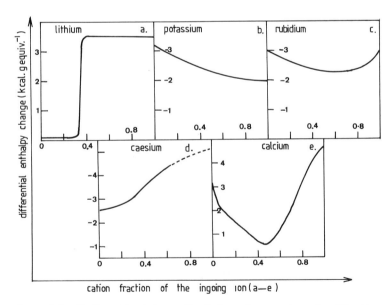

Figure 5.8 *Differential enthalpy profiles for ion exchange on the zeolite, (Na–A)* (Data from R. M. Barrer, L. V. C. Rees, and D. J. Ward, *Proc. R. Soc. London, A*, 1963, **273**, 180)

Ion Exchange Equilibria

Table 5.3 *Standard free energy* $\Delta G^{\ominus}$, *enthalpy* $\Delta H^{\ominus}$, *and entropy* $\Delta S^{\ominus}$ *changes accompanying some ion exchange reactions on gel strongly functional styrenic resins*

  a) *Alkali metal cations – Styrenesulfonate resin*
  b) *Alkaline earth cations – Styrenesulfonate resin*
  c) *Tetraalkylammonium cations – Styrenesulfonate resin*
  d) *Halide anions – benzyltrimethylammonium resin*

(Data from G. E. Boyd, 'Thermal Effects In Ion-Exchange Reactions With Organic Exchangers: Enthalpy and Heat Capacity Changes', in 'Ion Exchange In The Process Industries', Society of Chemical Industry, London, 1970, p. 261)

$RA + B \rightleftharpoons RB + A$ (at 25 °C) $A \quad B$		% DVB	$\Delta G^{\ominus}$ (kcal eq^{-1})	$\Delta H^{\ominus}$ (kcal eq^{-1}) (1 cal ≡ 4.184 Joules)	$\Delta S^{\ominus}$ (cal eq^{-1} K^{-1})
a) Na$^+$	Li$^+$	8	0.36	1.46	3.6
Na$^+$	H$^+$	8	0.23	1.18	3.0
Na$^+$	Na$^+$	8	0.00	0.00	0.00
Na$^+$	K$^+$	8	−0.26	−0.55	−1.0
Na$^+$	Cs$^+$	8	−0.33	−0.77	−1.5
b) Na$^+$	Mg^{2+}	8	−0.04	1.50	5.2
Na$^+$	Ca^{2+}	8	−0.35	1.41	5.9
Na$^+$	Sr^{2+}	8	−0.44	1.09	5.1
Na$^+$	Ba^{2+}	8	−0.79	0.75	5.2
c) Na$^+$	Me$_4$N$^+$	0.5	−0.28	−0.56	−1.0
Na$^+$	Me$_4$N$^+$	2.0	−0.09	−0.36	−0.9
Na$^+$	Me$_4$N$^+$	4.0	0.06	−0.19	−0.9
Na$^+$	Me$_4$N$^+$	8.0	0.27	0.04	−0.8
Na$^+$	Et$_4$N$^+$	0.5	−0.33	−0.50	0.2
Na$^+$	Pr$_4$N$^+$	0.5	−0.37	0.56	3.1
Na$^+$	Bu$_4$N$^+$	0.5	−0.46	2.21	9.0
d) F$^-$	Br$^-$	4	−1.56	−3.30	−5.8
Cl$^-$	Br$^-$	4	−0.57	−1.33	−2.5
Br$^-$	I$^-$	4	−0.88	−2.02	−3.8

for the observed effect upon selectivity, as measured by the selectivity coefficient, of such factors as:

1. equivalent ionic fraction loading of the resin.
2. resin crosslinking and water content

3. total exchange capacity
4. 'size' and charge of the counter-ions
5. nature of the fixed ion
6. nature of the solvent

In discussing selectivity behaviour the following convention is adopted. For ions A and B (cation or anion) the A–B system refers to exchange in the direction:

$$RA + B \rightleftharpoons RB + A \qquad (5.33)$$

where the resin is initially in the A form and the corrected mass action selectivity coefficient is denoted K_A^B, in appropriately defined concentration units: molar C, molal m, or equivalent ion fraction X. Alkali metal cation and halide anion exchange on styrenic strongly functional resins have been extensively studied largely because they are monovalent and exhibit well understood chemical periodicity. Where mechanistic theories of selectivity are concerned, multivalent and heterovalent ion exchange behaviour add enormously to the complexity of any model because of the varied ways in which multicharged counter-ions may be shared energetically among an equivalent array of monocharged fixed ions. Nevertheless studies of the energetics for monovalent ion exchange still enable projections to be made concerning ion exchange selectivity in general.

Figure 5.9 shows the trend most commonly encountered for binary exchange between alkali metal cations, and between alkali metal cation and hydrogen ion on styrenic sulfonic acid resins. The results depicted are part of an extensive study by Reichenberg and co-workers from which the following trends emerge for cation exchange in general:

1. Resin Loading. For the system (A–B), and $K_A^B > 1$, the corrected selectivity coefficient decreases with increasing $(X_B)_r$. Some exceptions to this pattern occur, for example the system H–Ag and some systems involving divalent ions.

2. Crosslinking. For simple inorganic cations, K_A^B usually decreases with decreasing resin crosslinking, but the opposite has been reported for alkylammonium cations which is discussed later when considering observed exchange energetics.

Ion Exchange Equilibria 121

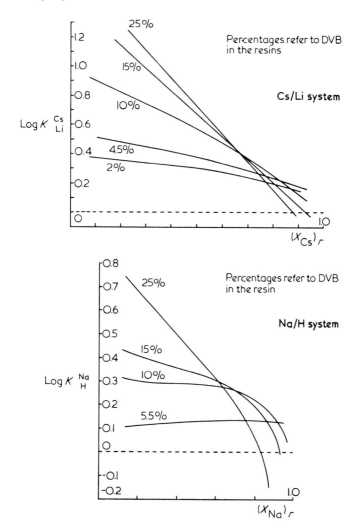

Figure 5.9 *Typical behaviour of the corrected selectivity coefficient with crosslinking and loading for monovalent cation exchange on styrenesulfonate resins*
(Reproduced by permission from D. Reichenberg, in 'Ion Exchange (A Series of Advances)', ed, J. A. Marinsky, Edward Arnold (London), Marcel Dekker (New York), 1966, Vol. 1, p. 227)

3. *Exchange Capacity.* The effect of exchange capacity is best described in terms of its influence upon the specific water content of a resin. Generally, the higher the water content per equivalent of

exchange capacity and therefore the swelling, the lower the selectivity coefficient.

4. Functional Group. The nature of the functional group can have a marked effect upon selectivity possibly causing not only a reversal for a pair of ions, but a reversal in affinity sequences. For example, dissociated carboxylate ($RCOO^-$) and phosphonate (RPO_3^{2-}) resins show affinities towards the alkali metal cations that are completely the reverse of that for sulfonate resins.

The dominant feature of the corrected selectivity coefficient behaviour depicted by Figure 5.9 is that there does not appear to be a *unique* ion affinity sequence for all values of crosslinking and ion loading. Clearly inversions, partial reversals ($K_A^B < 1$), and even total reversals [$K_A^B < 1$ for all $(X_B)_r$] occur for some systems. Within the limited scope allowed by an introduction to this most interesting of topics it must suffice to only briefly explain selectivity phenomena in terms of some of the proposed, and often controversial, mechanistic molecular theories. In attempting such a summary it is useful to express the free energy change ΔG_{ex} for an amount of exchange in terms of the changes in enthalpy ΔH and entropy ΔS in relation to the likely interactions occurring, for example:

$$\Delta G_{ex} = (\Delta G_{solv.} + \Delta G_{int.}) + \Pi(\Delta \bar{V}) \tag{5.34}$$

Where $\Delta G_{solv.}$ is the free energy change associated with changes in solvation of ions between the resin and external solution, $\Delta G_{int.}$ is the free energy change associated with specific interactions yet to be defined, and $\Pi(\Delta \bar{V})$ is the mechanical work (pressure–volume) term arising from the swelling pressure and difference in partial molar or molal volumes of the exchanging ions. It now follows that:

$$\Delta G_{ex} = [\Delta H_{is} + \Delta H_{ii} + \Delta H_{im}] - T[\Delta S_{is} + \Delta S_{ii} + \Delta S_{im}] + \Pi(\Delta \bar{V}) \tag{5.35}$$

Where is ≡ counter-ion and solvent interaction
ii ≡ counter-ion and fixed ion interaction
im ≡ counter-ion and resin matrix interaction

DILUTE SOLUTION CATION EXCHANGE

Mechanical Model

Matrix Volume Changes, $\Pi(\Delta \bar{V})$

This approach is important historically in that Gregor attributed selectivity to be entirely due to the mechanical work done during exchange against the swelling pressure Π of the matrix as given by the relationship:

$$RT \ln K_A^B = \Pi(z_B \bar{V}_A^* - z_A \bar{V}_B^*) \tag{5.36}$$

The terms $\bar{V}_A^*$ and $\bar{V}_B^*$ represent the partial molal volumes of the *solvated* counter-ions A and B respectively, and therefore for $K_A^B > 1$, the ion of smallest solvated volume is preferred by the resin phase. The description 'smallest solvated volume' is often equated to smallest hydrated radius for aqueous systems.

For simple cations this concept is found to hold and also predicts that the selectivity coefficient in the direction of preferred ion uptake increases with increasing resin crosslinking (greater swelling pressure Π). This simple model also explains the decrease in selectivity coefficient with increasing resin loading of the preferred ion $(X_B)_r$ since this is the direction of decreasing resin water content and hence reduction in swelling pressure. From a quantitative point of view the model lacks thermodynamic definition in that the concept of solvated ion volume is ill-defined and fails to explain selectivity inversions and reversals. To account for selectivity reversals in aqueous systems would require that the most highly hydrated ion could be spontaneously stripped, or partially stripped, of its hydration the most easily which is energetically unlikely.

Molecular Models

a) Fixed ion/counter-ion interaction, ΔH_{ii} and ΔS_{ii}

If selectivity were governed solely by interactions of a purely 'electrostatic' (coulombic) nature one could anticipate negative enthalpies and entropies of exchange due to dominant contributions by ΔH_{ii} and ΔS_{ii} respectively. This is indeed found for alkali metal cation exchange on styrenic sulfonate resins as illustrated by Table 5.3a.

Furthermore the decreasing affinity sequence, $Cs > K > Na > Li$ is in the order of increasing hydration of the ions or 'hydrated radius', which suggests a coulombic type interaction between counter-ions and the fixed ion based on 'ion size'. The simple model proposed by Pauley was based on this approach and identified affinity sequences for simple cations as being inversely related to their 'distance of closest approach' to the fixed anion. Harris and Rice proposed the formation of 'ion pairs' between the counter-ion and ionogenic group plus a further contribution to the resultant free energy of exchange from a configurational entropy change arising from the mutual repulsion of unpaired sites. This approach was extended by Katchalsky and others to cater for weakly crosslinked (highly swollen) resins where configurational entropy changes might be expected to be more significant.

The notion of 'hydrated ion size' being inversely related to selectivity holds quite well for simple inorganic cations but fails for some more complex inorganic cations and large organic cations. Also the thermodynamic energetics of exchange for the alkaline earth cations on sulfonic acid resins given in Table 5.3b show a dominant positive entropy contribution to the overall selectivity suggesting that effects other than pure electrostatic interactions are involved. Finally there remains the thorny problem of accounting for selectivity reversals with increased crosslinking and nature of the functional group.

b) Ion–solvent interactions, ΔG_{is}

A tidier picture emerges from the application to resins by Reichenberg of the work by Eisenman and Ling where solvation is not treated in terms of 'geometric ion size' but rather the solvation energetics as expressed by the free energy of solvation. This approach combined with an electrostatic interaction contribution gives the following expression:

$$\Delta G_A^B = e^2 \left(\frac{1}{r + r_A} - \frac{1}{r + r_B} \right) - (\Delta G'_A - \Delta G'_B) \qquad (5.37)$$

where e is the unit electron charge, r the crystallographic radius of the fixed anion, r_A and r_B the crystallographic radii of counter-ions A and B, and $\Delta G'_A$ and $\Delta G'_B$ the free energies of solvation of ions A and B respectively.

For aqueous systems and large r, the terms of the second bracket of equation 5.37 dominate and the free energy of exchange is governed by differences in the free energies of hydration of the counter-ions. Conversely, for small r, the terms of the first bracket in equation 5.37

become most significant and the selectivity is governed by electrostatic interactions. By considering different values of r, r_A, and r_B, the normal and reversed affinity sequences for alkali metal cation exchange on variously crosslinked styrenesulfonate resins are predicted, as are the reverse sequences found for the carboxylate and phosphonate ionogenic groups due to their high field strength.

The field strength of an ion is proportional to its charge but inversely proportional to its radius and is a measure of its polarizing power or how strongly it may be expected to interact with ions of opposite charge. The monovalent sulfonate ion is large and of low field strength such that for resins of low to moderate crosslinking counter-ion hydration energetics mainly control selectivity in that the normally most hydrated ion (with the greatest free energy of hydration) prefers the aqueous external phase. Increasing the crosslinking of a resin accentuates the preference for the normally most hydrated ion to seek solvation in the external solution and the selectivity for the preferred ion increases.

At very high crosslinking the functional groups become crowded which, through their close proximity, increases their field strength. Now the influence of the fixed anion is to compete with water for the solvation of a counter-ion even causing partial displacement of its hydration sheath in order to maximize mutual ion contact and therefore degree of interaction. It is this distortion of normal solvation energetics through the varying field strength of the ionogenic group that is believed to cause alkali metal cation selectivity reversals in styrenic sulfonate resins, and why ionogenic groups of high field strength such as carboxylate (COO^-) normally exhibit the reversed sequence for the same system. Energetically, ion exchange at sites of high field strength (high crosslinking) is preferred, and given the heterogeneous resin structure this could explain why the selectivity coefficient decreases with increasing uptake of the preferred ion, since remaining sites would be located in regions of lower crosslinking.

c) Hydrophobic hydration, ΔS_{is}

As described by Franks, the aqueous dissolution of non-polar solutes or more polar solutes carrying a non-polar substituent often results in a large negative excess entropy and negative partial molar volume of mixing. Ideally the entropy of mixing should be positive, but it is thought that the presence of non-polar species causes the normal hydrogen bonded structure of water to become enhanced and more ordered (entropy decrease), creating 'cavities' which accommodate the non-polar species whose hydration is now through a weak interaction with 'free' or loosely structured water. This phenomenon has been termed *hydrophobic hydration* by Franks and can be brought about by organic ions as well as non-polar solutes.

The enhancement of water structure by organic ions is believed to be the driving force behind the high affinity of tetraalkylammonium cations over conventional cations on styrenesulfonate cation exchange resins. In such systems, the selectivity for the organic cation increases with increasing size of the alkyl group and decreasing crosslinking of the resin. As can be seen from Table 5.3c the spontaneous decrease in free energy of the system ($K_A^B > 1$) arises largely from the large positive entropy change ΔS_{is} occurring through the preferential uptake or *hydrophobic bonding* of the organic cation in the resin phase, and a small increasingly positive enthalpy change due to the 'melting' of external water structure. For even larger organic cations, Feitelson argues that water structure enforced hydrophobic bonding (positive ΔS) may be overshadowed by large negative enthalpies and negative entropies of exchange due to strong van der Waals interactions between the ion's non-polar residues and the resin matrix (ΔH_{im} and ΔS_{im}). Thus, as Figure 5.10 shows, strong selectivity towards the

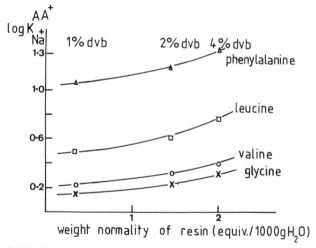

Figure 5.10 *Selectivity coefficient behaviour for amino acid cation exchange against sodium on styrenesulfonate resins of various percent crosslinking*
(Reprinted from J. Feitelson, in 'Ion Exchange (A Series of Advances)', ed. J. A. Marinsky, Marcel Dekker, New York, Vol. 2, p. 135, by courtesy of Marcel Dekker Inc.)

organic cation is observed but which now increases with ion size and degree of crosslinking, always providing ion exclusion effects remain absent.

DILUTE SOLUTION ANION EXCHANGE

The selectivity coefficient behaviour for anion exchange is somewhat less systematic compared with cation exchange; and as Figure 5.11 shows, opposite behaviour can occur with increasing loading of the resin together with inversions and reversals. Unlike cations, with the exception of possibly small ions such as fluoride, chloride, and hydroxide, most anions are much less solvated than cations of the same charge.

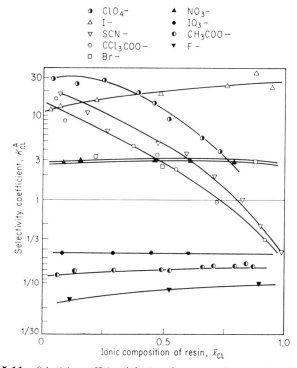

Figure 5.11 *Selectivity coefficient behaviour for anion exchange against chloride ion on a quaternary ammonium (Type 2) strongly basic resin*
(Reproduced by permission from F. Helfferich, 'Ion Exchange', McGraw–Hill, London and New York, 1962)

Mechanical Model, $\Pi\Delta\bar{V}$

Solvated ion volumes alone, and therefore pressure–volume considerations, fail to account for observed ion affinity sequences even for the simple halide ions.

Molecular Models

a) Fixed ion/counter-ion interactions, ΔH_{ii} *and* ΔS_{ii}

As indicated in Table 5.3d, significant negative enthalpies and entropies of exchange for the exchange of halide ions on quaternary ammonium resins could be interpreted in terms of pure electrostatic interactions. However such a coulombic model invoking solvated ion size or 'distance of closest approach' fails to conform to the observed affinity order: $I^- > Br^- > Cl^- > F^-$. If one includes a *larger* polyatomic anion such as perchlorate, $Cl^{VII}O_4^-$ the experimental sequence is $Cl^{VII}O_4^- > I^- > Br^- > Cl^- > F^-$ which clearly contradicts any simple coulombic attraction theory.

b) Ion–solvent interactions, ΔG_{is}

As for cation exchange described previously, the net reduction in free energy of the system is better explained in terms of the net negative enthalpy contribution (ΔH_{is}) arising from the naturally most solvated (hydrated) ion A preferring the external aqueous phase, plus the ionogenic group/counter-ion (B) interaction. The model proposed by Eisenman and Ling for cation exchange can, in principle, be adapted to apply to anion exchange by taking into account the further lowering of free energy arising from the field strength of the ionogenic group in promoting close approach of the counter-ion.

c) Water structure enforced ion pairing, ΔS_{is}

The debate and controversy concerning the origins of anion exchange selectivity continues to be fuelled through a very convincing theory proposed by Diamond and his co-workers. For polyatomic ions of high field strength (strongly basic), the stronger they are, the more likely the are to bond to water and therefore prefer the aqueous phase. In other words for hydrophilic anions of similar size the weaker its parent (conjugate) acid the more it should prefer the external aqueous solution.

Also large lowly charged anions of low field strength enhance the external hydrogen bonded water structure such as to resist the intrusion by the large ion and thereby force the ion into the resin phase (ΔS_{is} positive), where the hydrophobic polymer matrix and presence of relatively unstructured water promotes 'pairing' or interaction with the fixed cation. This mechanism is termed 'water structure enforced ion pairing' by Diamond and explains, for example, the affinity sequence in order of decreasing ion size and increasing base strength for the series:

$$Au^{III}Cl_4^- > Cl^{VII}O_4^- > I^- > Br^- > Cl^- > F^-$$

For ions of similar base strength the resin will prefer the larger ion because of its stronger enhancement of water structure in the external aqueous dilute solution. This is particularly noticeable for organic anions of increasing molecular weight or, in the case of large pendant organic substituents, possible additional van der Waals bonding as advocated by Feitelson for organic cation exchange. An interesting affinity sequence is afforded by the series:

$$\text{methanoate} > \text{ethanoate} < \text{butanoate}$$

On the grounds of size, methanoate and ethanoate affinities appear misplaced, but methanoate is a weaker base (stronger parent acid) than ethanoate, and therefore is less hydrated than ethanoate thus preferring the resin phase. Water structure enforced ion pairing has been proposed as the reason for significant anion exchange affinities ($K_A^B > 1$) even in resins of extremely low crosslinking. Ion exchange selectivity may be broadly considered as a competition for solvation between phases of widely differing hydrophilic character, and therefore it is both predictable, and verifiable, that affinity sequences can be altered greatly in non-aqueous and mixed aqueous–organic solvents for both cations and anions.

d) Matrix charge separation

It will have been noticed that the field strength of the ionogenic group plays an important part in current selectivity theories which, when fully understood, should provide a degree of unification of the present similar but fundamentally opposed ideas. This view is no better supported than by recent work of Clifford and Weber who show that the greater the charge separation (distance between ionogenic groups) within a resin matrix the greater the affinity for monovalent over divalent ions for ion exchange generally on styrenic and acrylic resins (Figure 5.12).

Therefore charge separation and enhancement of selectivity towards monovalent ions may be brought about by incorporating the ionogenic group as pendant groups rather than in the polymer chain, increasing the size of the ionogenic group, and by high crosslinking to inhibit configurational entropy changes, thus preventing a divalent ion from sharing its charge optimally between two fixed ions. An important example of the 'charge separation' theory is seen with the aqueous tertiary (chloride, nitrate, sulfate) system, where the benzyltriethylammonium strongly functional resin in the chloride form prefers nitrate over sulfate, as expressed by the separation factor $\alpha_{NO_3^-}^{SO_4^{2-}}$, which is opposite to the affinity sequence found for the common and smaller benzyltrimethylammonium functional group. The nitrate

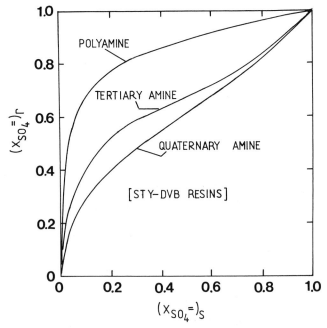

Figure 5.12 *The effect of amine functionality on the selectivity of SO_4^{2-} over NO_3^- for styrene–divinylbenzene anion exchangers. [Note: The matrix charge separation increases in the order: polyamine < tertiary amine < quaternary amine]*

(Reproduced by permission from D. Clifford and W. J. Weber, *Reactive Polymers*, 1983, **1**, 77)

over sulfate selectivity is enhanced with increasing size of the tertiary alkylamine in the order:

$$\text{tributyl} > \text{tripropyl} > \text{triethyl} > \text{trimethyl}$$

SELECTIVITY IN CONCENTRATED SOLUTIONS

Ion exchange between resins and concentrated external electrolytes is generally far more complex compared with the situation for dilute solution. The same general selectivity determining considerations apply, but deviations from behaviour described for dilute systems can be anticipated on the basis of:

a) Lower water activity in the solution phase may result in the full solvation requirements of ions not being met, thus changing the ion–ion and ion–solvent energetics described for dilute solutions thereby changing the relative affinities.

b) The high external concentration of ions results in 'ionic interactions' playing a larger role in determining selectivity.

c) Theoretical treatments become much more complex through changes in resin water content and significant co-ion uptake by the resin. Therefore any thermodynamic or Donnan equilibrium theory must now take account of changes in water activity and consequently swelling pressure.

d) Often, complex ion formation in concentrated electrolytes will totally change the charge on ions taking part in the exchange. A practical example of this is afforded by iron(III) and zinc(II) cations which in concentrated hydrochloric acid solution form the anionic complexes $Fe^{III}Cl_4^-$ and $Zn^{II}Cl_4^{2-}$ which are strongly exchanged on chloride form strongly basic anion exchange resins, thus effecting an easy separation from other cations present (see Box 5.2 and Chapter 8).

e) Ligands such as ammonia, amines, and polyhydric alcohols may be exchanged between an external aqueous phase and resins carrying ions capable of forming coordination complexes, thus providing a powerful technique for studying complex ion structure and complex formation equilibria.

BOX 5.2   Sorption of a Cation as an Anionic Complex

A typical impurity of commercial concentrated hydrochloric acid is iron. This is present as an anionic complex $FeCl_4^-$. This is removed on a strong base anion exchange resin in the chloride form. As the complex only exists in strong acid solution, it is easily removed from the resin by displacing the acid by water and thus diluting the acidic environment.

Requirements

5 ml of SBA resin (Cl form)
Concentrated HCl in which iron is dissolved to give a 0.005 M Fe solution
Narrow glass column, 12 mm diameter (approx.)

Procedure

The resin is put into the glass column and a nylon wool plug placed on top of the resin to prevent the resin from floating in the acid.
Pour the acid through the resin at a flow rate of about 20 ml per hour. The resin can treat about 1 litre of acid if required.
Finally rinse the column slowly with deionized water when the iron will be displaced as a narrow coloured band.

TIME REQUIRED – 1 hour 30 minutes

In general the complicating factors described above, and 'electroselectivity' effects, make equilibrium behaviour in concentrated electrolytes difficult to predict. However some success has been achieved in modelling selectivity coefficient behaviour for simple systems.

FURTHER READING

F. Helfferich, 'Ion Exchange', McGraw–Hill, London and New York, 1962.

D. Reichenberg, 'Ion-Exchange Selectivity', in 'Ion Exchange', (A Series of Advances)', ed. J. A. Marinsky, Edward Arnold (London), Marcel Dekker (New York), Vol. 1, 1966, p. 227.

J. Feitelson, 'Interactions Between Organic Ions and Ion Exchange Resins', in 'Ion Exchange, (A Series of Advances)', ed. J. A. Marinsky, Marcel Dekker, New York, Vol. 2, 1969, p. 135.

K. Dorfner, 'Ion Exchangers', ed. K. Dorfner, Walter de Gruyter, Berlin and New York, 1991.

F. Franks, 'Aqueous Solutions of Electrolytes', in 'Water', Royal Society of Chemistry, London, 1983, Ch. 10, p. 57–68.

W. J. Brignal, A. K. Gupta, and M. Streat, 'Representation and Interpretation of Multicomponent Ion Exchange Equilibrium', in

'The Theory and Practice of Ion Exchange', Society of Chemical Industry, 1976, p. 11.1.

D. Clifford and W. J. Weber, 'The Determinants of Divalent/Monovalent Selectivity in Anion Exchangers', *Reactive Polymers*, 1983, **1**, 77.

R. M. Diamond and D. C. Whitney, 'Resin Selectivity In Dilute to Concentrated Aqueous Solutions', in 'Ion Exchange (a Series of Advances)', ed. J. A. Marinsky, Edward Arnold (London), Marcel Dekker (New York), Vol. 1, 1966, p. 277.

M. Abe, 'Ion Exchange Selectivities of Inorganic Ion Exchangers', in 'Ion Exchange Processes: Advances and Applications', ed. A. Dyer, M. J. Hudson, and P. A. Williams, Special Publication No. 122, Royal Society of Chemistry, Cambridge, 1993, p. 199–213.

Chapter 6

The Kinetics and Mechanism of Ion Exchange

BASIC CONCEPTS

The rate at which an ion exchange reaction proceeds is a complex function of several physico-chemical processes such that the overall reaction rate may be influenced by the separate or combined effects of:

1. Concentration gradients in both phases
2. Electrical charge gradients in both phases
3. Ionic interactions in either phase
4. Exchanger properties (structure, functional group)
5. Chemical reactions in either phase

As yet, no analytical and readily integrated unique mathematical function of the type $-d\bar{C}_A/dt = f(\bar{C}_A)$ exists for describing the kinetics, where $(\bar{C}_A)$ is the resin phase concentration of the counter-ion A initially in the exchanger, B the ion in solution, and t the elapsed time. However, analytical solutions of the rate equations are available which account for the observed rate behaviour under specified circumstances or boundary conditions.

Studies of ion exchange reactions on organic exchangers have identified the possible rate controlling steps to be:

1. Coupled diffusion or transport of counter-ions in the 'external' solution phase.
2. Coupled diffusion or transport of counter-ions within the ion exchange resin.
3. Chemical reaction at the sites of the functional groups within the exchanger.

An understanding of the kinetics of ion exchange reactions has application in two broad areas. Firstly, it helps to elucidate the nature of the various fundamental ionic transport mechanisms which control or contribute to the overall exchange rate. Secondly derived numerical parameters such as 'rate constants', mass transfer coefficients, or diffusion coefficients found from a rate investigation are of value when making projections concerning the dynamic behaviour of columns and in process design.

This second area of application is a chemical engineering one where practical situations invariably involve changing constraints or boundary conditions. The time-dependent coupling of mass and charge transfer which epitomizes the ion exchange situation is extremely complex, and it can be solved only by highly computerized calculations employing a sophisticated model. Fortunately, in many practical situations, a rigorous complex numerical kinetic analysis may often be substituted by a more simplified approach giving rate parameters still sufficiently accurate for process design and performance prediction purposes.

Before considering some analytical and numerical functions found to describe the kinetics of ion exchange it is useful to consider the various rate controlling steps in more qualitative terms. Figure 6.1 represents schematically the three fundamental rate determining mechanisms put forward as controlling the overall rate of exchange of say ion B in solution with beads of resin RA in a well stirred suspension.

1. Coupled Diffusion of Counter-ions in the 'External' Solution

Efficient stirring of the resin with the solution ensures the elimination of ion concentration gradients in the bulk solution such that mass transfer in this phase is purely by convection and not rate determining. However, it is a fact that convective mass transfer diminishes at close proximity to the resin bead surface, and although hydrodynamically ill-defined, a stagnant liquid layer or *film* may be considered to surround the exchanger particles across which ion mass transfer is controlled by planar, one dimensional diffusion. The need to preserve electroneutrality during ion exchange, by whatever mechanism, requires that an equal and opposite counter-ion flux must always apply. Therefore rate control by mass transfer in the 'external' solution is interpreted as *coupled mass transfer* across the hypothetical film or *Nernst layer* surrounding the resin particles by a mechanism of diffusion called *film diffusion*.

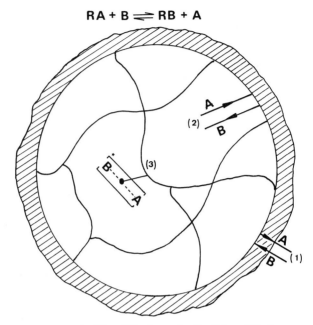

1. Film diffusion ; 2. Particle diffusion ;
3. Chemical reaction

Figure 6.1 *Rate determining steps in ion exchange (schematic)*

The driving force for mass transfer is the concentration difference or, more correctly, the concentration gradient of counter-ions between the two boundaries of the film. If theoretical refinements are ignored for the time being, the momentary flux equation for ion A may be written in terms of Fick's first law:

$$J_A = \frac{D_A \Delta C_A}{\delta} \qquad (6.1)$$

Where J_A is the flux (kmol m^{-2} s^{-1}) of ion A under a finite concentration difference ΔC_A (kmol m^{-3}) across the film of thickness δ (m). Flux is a vector quantity (directional) and an equivalent condition may be expressed for the diffusion of ion B in the opposite direction. As will be realized later when considering rate equations there is a requirement for the diffusion of ions A and B to be coupled electrically and this requires the momentary flux equations to contain a single 'average value' of the diffusion coefficient rather than the separate diffusion coefficients D_A and D_B (m^2 s^{-1}). The diffusion coefficient may be viewed as the flux per unit concentration gradient, the values of which for film diffusion are not vastly different from those found in bulk aqueous solutions and show activation energies of about 10–20 kJ eq^{-1}. The electroneutrality constraint with regard to mass and charge transfer across the film presupposes that the conditions represented by equations 6.2 and 6.3 are met:

$$z_A J_A + z_B J_B = 0 \quad \text{(zero net charge)} \quad (6.2)$$

and

$$z_A C_A + z_B C_B = C \quad \text{(electroneutrality)} \quad (6.3)$$

where z_A, z_B are the electrovalencies of the counter-ions, and C is the total external concentration (keq m^{-3}). The above stipulations imposed upon the time dependent diffusional driving force is the start point for establishing diffusion rate equations which are then further manipulated to take account of particular boundary conditions and theoretical refinements.

2. Coupled Diffusion of Counter-ions in the Resin

Mass transfer in the film and resin particle are sequential processes and either process may be rate controlling, *i.e.* the slowest step. By *simplified* analogy with the previous case for film diffusion, the average flux condition for transfer of ion A in the resin phase may be written:

$$\bar{J}_A = \frac{\bar{D} \Delta \bar{C}_A}{r_o} \quad (6.4)$$

Now, the driving force is the concentration gradient between the interior of the resin and the resin–solution interface for ion A and resin beads of radius r_o. The bar notation represents the resin phase and $\bar{D}$, again, an 'average' diffusion coefficient for ions A and B within the resin. Resin phase diffusion coefficients are about one or two orders of magnitude smaller than found for aqueous solutions because of the steric resistance offered by the copolymer matrix.

If mass transfer in the resin determines the overall rate of ion exchange then the reaction is said to be *particle* or *intraparticle diffusion* controlled. The simplest models for particle diffusion control regard the resin as a homogeneous gel phase for which Mackie and Mears proposed the following relationship between internal ($\bar{D}_i$) and external (D_i) diffusion coefficients and the internal porosity, or void volume, ε of the exchanger:

$$\bar{D}_i = D_i \left[\frac{\varepsilon}{(2-\varepsilon)}\right]^2 \quad (6.5)$$

For practical purposes ε may be equated to the weight fraction of solvent (water) in the resin whereby equation 6.5 has been found to hold quite well for simple inorganic ions and swollen *gel* resins. As might be expected, deviations occur for counter-ions showing strong interaction with the matrix or the ionogenic group, and also for macroporous resins where differing diffusion resistances of the gel microstructure and solvent filled macropores combine to alter the nature of the intraparticle diffusion characteristics. Besides the purely steric barrier to ion diffusion arising from the copolymer structure, the matrix charge distribution along a diffusion path intuitively suggests that diffusing counter-ions would be retarded by their interaction with a periodic and varying electrostatic force field. Therefore selectivity considerations might be expected to influence the nature of the rate controlling mechanism, which is indeed the case.

3. Chemical Reaction Rate Control

True chemical reaction at the sites of the functional groups is represented purely schematically in Figure 6.1 by an imaginary transition state complex between ions A, B, and the ionogenic group. Such a concept involves the making and breaking of ionic, covalent, or dative bonds which has never been shown unequivocally to be solely rate controlling for the exchange of simple aqueous ions on organic exchangers. Reactions between simple, freely dissociated, aqueous ions are usually very fast and therefore not rate controlling, but some published data suggest chemical reaction rate control for the exchange of transition metal ions or complex ions capable of strong chelate type complex formation with iminodiacetate or phosphonate functional groups. Generally the evidence for truly chemical reaction rate control is inconclusive, but to acknowledge a coupled 'reaction–diffusion' mechanism for some systems is somewhat more acceptable and in line with the reported higher activation energies (60–100 kJ eq^{-1}) for particle diffusion in chelating resins, compared with about 25–40 kJ eq^{-1} for intraparticle diffusion of simple ions in conventional resins.

Examples of accompanying chemical reactions influencing film and intraparticle diffusion rate control is afforded by instances where a counter-ion can react with the functional group or with the co-ion, for example:

1. neutralization of strong functional groups:

$$RSO_3^-H^+ + NaOH \rightarrow RSO_3^-Na^+ + H_2O$$

2. complex formation with a co-ion:

$$(RSO_3^-)_4Ni_2^{2+} + Na_4EDTA \rightarrow 4\,RSO_3^-Na^+ + Ni_2EDTA$$

3. association of weakly functional groups:

$$RCOO^-Na^+ + HCl \rightarrow RCOOH + NaCl$$
$$RNH_3^+Cl^- + NaOH \rightarrow RNH_2 + NaCl + H_2O$$

4. dissociation of weakly functional groups:

$$RCOOH + NaOH \rightarrow RCOO^-Na^+ + H_2O$$
$$RNH_2 + HCl \rightarrow RNH_3^+Cl^-$$

For these and similar systems the original source, resin or solution, of the counter-ion being chemically consumed and the nature of the co-ion greatly influence the observed kinetics. The association–dissociation of weakly functional resins is of particular practical interest since in these instances a reactive and non-reactive core respectively forms within the resin which shrinks towards the bead centre as exchange proceeds. This 'Shrinking Core' or 'Shell Progressive' mechanism is usually particle diffusion controlled and explains why exchange on weakly functional resins is invariably flow-rate sensitive under column operation.

BOX 6.1 An Experiment to Show 'Moving Boundary' Formation during Particle Diffusion Accompanied by Chemical Reaction

This experiment is described in detail by Dorfner[1] and demonstrates the shrinking core mechanism, observable as diffusion of an entering ion within the resin as it undergoes exchange accompanied by chemical reaction.

Requirements

Premium gel SAC resin, 8–10% DVB and ideally of bead size 0.7–1.3 mm diameter
Small column (~ 12 mm diameter)
2 N H_2SO_4 solution
2 N NaCl solution
2 N $(NH_4)_2SO_4$ solution
Ammoniacal $CuSO_4$ solution (0.25 M $CuSO_4$/2.5 M NH_4OH)
Watchglass
Microscope
Filter crucible and Buchner filter assembly

Procedure

Prepare a small ion exchange column using 10 cm³ of SAC resin (H or Na form). The resin is converted to the ammonium form by passing 2 BV of 2 N ammonium sulfate solution slowly through the bed followed by a rinse with deionized water. Two bed-volumes of ammoniacal copper sulfate solution (0.25 M $CuSO_4$/2.5 M NH_4OH) is passed through the column which converts the resin to the intense blue copper(II) tetrammine form, $R_2^{2-}\ Cu(NH_3)_4^{2+}$. The resin is rinsed and stored in a tightly stoppered bottle.

A shallow layer of the cuprammonium form resin is placed in the filter crucible connected to a Buchner vacuum filtration assembly. The resin is briefly contacted with a small quantity of 2.5 M (5 N) sulfuric acid which initiates elution of the cuprammonium ion. The acid is filtered off and the resin immediately rinsed with excess deionized water to remove all traces of acid. A sample of wetted beads are viewed under a microscope using transmitted light; whereupon an inner unreacted blue core is seen to be surrounded by a transparent outer zone. This experiment is also a demonstration of a cation (Cu^{2+}) being sorbed onto a resin as a cationic complex with the ion originally present on the resin (NH_4^+).

[1]K. Dorfner, 'Laboratory Experiments and Education in Ion Exchange', in 'Ion Exchangers', ed. K. Dorfner, Walter de Gruyter, Berlin and New York, 1991, Chapter 1.2, p. 425.

RATE EQUATIONS

A rate equation is simply a mathematical function that is found to describe the observed rate of release or uptake of a given counter-ion with respect to the resin or external solution. This is not to be confused with the rate mechanism, or to put it another way, a fit of data to a rate equation does not necessarily prove the mechanism. Many types of mathematical functions have been shown to describe ion exchange rate data, but further experimental tests are required to establish the mechanism.

Also, all other considerations being equal, since intraparticle diffusion coefficients are very much smaller than for film diffusion one might expect the rate of exchange always to be particle diffusion controlled. However, all considerations are not equal since the concentration driving force for film diffusion ($C_A{}^* - C_A$) and particle diffusion ($\bar{C}_A - \bar{C}_A{}^*$) are different. $C_A{}^*$ and $\bar{C}_A{}^*$ are the solution side and resin side concentrations of ion A at the exchanger surface–solution interface which are assumed to attain instantaneous equilibrium at all times during an ion exchange reaction. Resistance to mass transfer at the resin surface is negligible whilst the resin surface is clean, but should the bead surface become fouled in some way then not only would this seriously impede the rate of exchange but also normal equilibrium behaviour may not be attained. In column operations this effect gives rise to the phenomenon of *kinetic leakage* discussed later in Chapter 7.

Analogy with Chemical Kinetics

An important early application of ion exchange was in water softening and this particular system has been studied in some detail. The dynamics of ion exchange column behaviour was studied by using chemical kinetic models as a basis for interpreting the data. Basic laws of chemical kinetics were found to apply in a mathematical sense to the rate data from ion exchange reactions, but the mechanism of ion exchange is now known not to be one of chemical reaction at the exchange sites.

Rate laws of the type which describe bimolecular second order chemical reactions might be expected to be a model for ion exchange reactions, and indeed this was the case for exchangers of both natural and synthetic origin. For example, the rate of ion exchange could be described by a bimolecular second order rate equation for irreversible reaction of the form:

$$\frac{dX}{dt} = k_2(a - X)(b - X) \quad (6.6)$$

In its integrated form equation 6.6 may be written:

$$\log\left[\frac{b(a - X)}{a(b - X)}\right] = \frac{(a - b)}{2.303} k_2 t \quad (6.7)$$

where b represents the initial concentration of counter-ions in the external solution, and a refers to the concentration of ions originating from the exchanger which are regarded as dissolved in the chemical reactant sense. The coefficient k_2 is the second order rate constant and X represents the external ionic concentration of the species originally in the exchanger at a given time t.

Equation 6.7 was useful in interpreting the rate and activation energy data from early batch experiments carried out in a stirred vessel. For simple dilute aqueous ion exchange systems the transport of ions up to and away from the surface of the exchanger (film diffusion) was concluded to be the rate determining step by assuming that any kind of 'diffusional' control within the exchanger would be reflected by a much greater temperature coefficient of reaction rate and therefore activation energy. Later studies have shown that this conclusion is often, but not always, true. Further early work showed that, although ion exchange involved a heterogeneous system, the exchanger phase may be regarded as a fully dissolved reactant and the laws of homogeneous chemical kinetics may be usefully applied, regardless of the true mechanism. Reaction rate studies along the predescribed lines were typical of early investigations into the kinetics of ion exchange reactions.

Two further examples are worthy of comment. Firstly the second order bimolecular rate equation for reversible chemical equilibrium given by equation 6.8 was found to fit rate data for ion exchange on carbonaceous exchangers:

$$\log\left[\frac{ab(X_e + X) - X_e(a + b)X}{ab(X_e - X)}\right] = \frac{[2ab - (a + b)X_e]k_2 t}{2.303 X_e} \quad (6.8)$$

Secondly, unimolecular first order kinetics given by equation 6.9 is found to fit ion exchange rate data generated under film diffusion control.

$$\log\left[\frac{X_e}{X_e - X}\right] = \frac{akt}{2.303 X_e} \quad (6.9)$$

The various symbols have the same meaning as in equation 6.7; X_e is the equilibrium concentration in solution of the ion originally in the exchanger, and k represents the reaction rate constant for the chosen direction of exchange. The fractional attainment of equilibrium is given by X/X_e whilst the fractional equilibrium conversion of the resin is equal to X_e/a.

The rate model represented by equation 6.9 is particularly interesting since it will be shown shortly that a similar function arises from a consideration of formal diffusion theory. Therefore, providing it is established by experiment that the pseudo 'rate constant' is truly constant over the range of experimental boundary conditions employed (a, b, X_e) it remains perfectly valid to equate its value to appropriate mass transfer parameters required by diffusion theory.

By way of example, Figure 6.2 shows a first order rate plot for the stirred contact of sodium form styrenesulfonate cation exchange resin (12% DVB) with 0.001 M hydrochloric acid solution at 25 °C. The system data and calculated 'rate' constant are given in Table 6.1. The activation energy may be found from the temperature dependence of the rate constant and was found to equal 16.7 kJ eq^{-1}. This same data is redeployed later according to more rigorous diffusion theory.

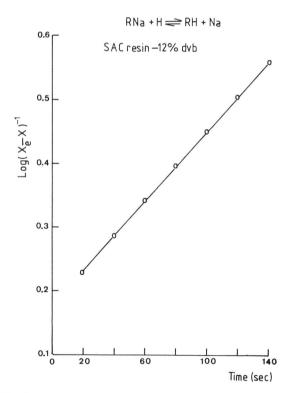

Figure 6.2 *Interpretation of ion exchange rate measurements according to first order chemical kinetics (equation 6.9)*
(Data from C. E. Harland, Ph.D. Thesis, Department of Mining and Mineral Sciences, University of Leeds, 1972)

Table 6.1 System parameters and derived 'rate data' for sodium–hydrogen exchange kinetics on a styrene sulfonate resin (12% DVB) – finite volume. (Data from C. E. Harland, Ph.D. Thesis, University of Leeds, 1972)

System data	Chemical model	Film diffusion models	
	Equation 6.9 (Figure 6.2)	Equation 6.20 (Finite Volume)	Equation 6.21 (Figure 6.3)
$RNa + H \rightleftharpoons RH + Na$ Resin: SAC (12% DVB)			
$\bar{C} = 4.5$ keq m$_r^{-3}$	$a = 2.475$ eq m^{-3}		
$C = 1.0 \times 10^{-3}$ keq m^{-3}	$b = 1.0$ eq m^{-3}		
$V = 2.75 \times 10^{-4}$ m^3	$X_e = 0.676$ eq m^{-3}		
$\bar{V} = 1.53 \times 10^{-7}$ m^3	$Slope = 2.71 \times 10^{-3}$ sec^{-1}	$Slope = 2.71 \times 10^{-3}$ sec^{-1}	$Slope = 2.2 \times 10^{-3}$ sec^{-1}
$r_o = 1.5 \times 10^{-4}$ m	$k = 7.25 \times 10^{-4}$ sec^{-1}	$D = 4.0 \times 10^{-9}$ m^2 sec^{-1}	$D = 1.5 \times 10^{-8}$ m^2 sec^{-1}
$\delta = 10^{-5}$ m			$(\bar{X}_{Na})_e = 0.73$
Temperature = 24.9 °C			$\alpha_H^{Na} = 1.3$
Stirring speed = 930 rpm			

Thus it is seen that the validity of a chemical kinetic model does extend as far as providing a linear plot from which a mass transfer related rate constant may be obtained. Further literal chemical kinetic interpretation breaks down for diffusion controlled ion exchange reactions as evident from observing that:

1. The 'rate constant' is not constant for all values of a, b, X_e, etc.
2. The observed rate of exchange is not always dependent upon the external concentration as required by a chemical rate controlling mechanism.
3. The 'rate constant' varies with particle size of the exchanger.
4. 'Rate constants' for forward and back exchange bear no relationship to the actual value of the selectivity coefficient for the system.
5. A 'rate constant' is sometimes found to be dependent upon the degree of agitation (stirring rate).

Thus it follows that any or all of the criteria 1–5 above would be a basis of discerning between diffusion and true chemical reaction as rate controlling steps.

Diffusion Kinetics

It was originally by experiment, rather than through any specific theoretical requirement, that the overall rate of ion exchange was found to be governed by a diffusion mechanism. It remained for Boyd and his co-workers as far back as 1947 to establish model rate equations by which one could describe, and discern between, film and intraparticle diffusion control. The fundamentals of Boyd's treatment still stand but the all-so-important later inclusion of an electric field gradient introduced by Helfferich in 1956 has had a profound influence on subsequent theoretical developments. The application of the most advanced and mathematically rigorous theoretical models to actual systems is extremely difficult demanding extensive computerized numerical analysis. Fortunately the sensible use of more simplified approaches serve very well to both identify the rate controlling mechanism, and to generate 'rate parameters' for practical use.

Particle Diffusion (Ideal)

Given that the fluxes of the two counter-ions A and B are rigidly coupled through the requirements of equivalent exchange and preservation of electrical neutrality, then assuming Fick's First Law of diffusion the following flux condition applies:

$$\bar{J}_A = -\bar{D} \cdot \text{grad}\, \bar{C}_A \tag{6.10}$$

where grad $\bar{C}_A$ is the concentration gradient of ion A across an element of the exchanger.

The coupling of the flux and the time dependence of the counter-ion concentration is achieved through the material balance condition and Fick's Second Law (often termed the condition of continuity). For spherical particles and a *constant interdiffusion coefficient*, $\bar{D}$, the foregoing considerations give the following partial differential equation for particle diffusion kinetics:

$$\frac{\partial \bar{C}_A}{\partial t} = \bar{D}\left[\frac{\partial^2 \bar{C}_A}{\partial r^2} + \frac{2}{r}\left|\frac{\partial \bar{C}_A}{\partial r}\right|\right] \tag{6.11}$$

where r is the distance travelled from the centre of the particle. Equation 6.11 may be solved analytically under appropriate boundary conditions to give the following rate function:

$$F(t) = 1 - \frac{6}{\pi^2}\sum_{n=1}^{n=\infty}\frac{1}{n^2}\cdot \exp\left[-\frac{\bar{D}\pi^2 n^2 t}{r_o^2}\right] \tag{6.12}$$

where $F(t)$ is the degree of conversion of the resin, or more correctly, the fractional attainment of equilibrium at time t, r_o is the radius of the exchanger particle, and $\bar{D}$ the diffusion coefficient in the exchanger.

The boundary condition applying to the solution given by equation 6.12 is that for a resin initially in the A form the concentration of A in the external solution remains zero at all times. Boyd and his co-workers achieved this by using radio-tracer techniques in 'shallow-bed' experiments and followed the *isotopic exchange* in a system which was otherwise at equilibrium. Alternatively, the appropriate boundary condition may be closely approached in conventional stirred reactor experiments if the volume of the external solution is large, and the ingoing ion B is a microcomponent of the system. Thus this particular solution for particle diffusion kinetics is often termed the *unlimited bath* or *infinite volume* case and only applies strictly to isotopic exchange where the interdiffusion coefficient is constant and there is *no selectivity* ($K_B^A = 1$).

Tables of the function $(\bar{D}t/r_o^2)$ as a function of F are available in the literature which greatly facilitates testing the fit of rate data to this particular particle diffusion model. For very low values of F, equation 6.12 reduces to the well known square root law for solid phase diffusion:

$$F(t) = \frac{6}{r_o} \left| \frac{\bar{D}t}{\pi} \right|^{1/2} \quad (6.13)$$

The mathematical solution to equation 6.10 is also available for the more general case of 'finite volume' or 'limited bath' conditions, where equilibrium at the particle–solution interface is assumed at all times and the macro-concentration of ions A in the external solution is time-dependent. An analogous situation arises in the theory of heat transfer where the mathematical solution also serves the case of ion exchange:

$$F = \frac{W+1}{W}\left\{1 - \frac{1}{\alpha - \beta}[\alpha \exp \lambda - \beta \exp \lambda']\right\} \quad (6.14)$$

where W is the ratio of the total equivalent concentrations in the exchanger and the external solution. The terms λ and λ' represent the functions

$$\frac{\alpha^2 \bar{D}t}{r_o^2}\left[1 + \mathrm{erf}\,\alpha\left(\frac{\bar{D}t}{r_o^2}\right)^{1/2}\right]$$

and

$$\left[1 + \mathrm{erf}\,\beta\left(\frac{\bar{D}t}{r_o^2}\right)^{1/2}\right]\beta^2 \frac{\bar{D}t}{r_o^2}$$

respectively.* The coefficients α and β are the roots of the quadratic equation $(X^2 + 3WX - 3W = 0)$. As for the previous case, the function has been tabulated for the convenience of application.

In general, the observed rates of isotopic exchange agree well with theory in that the derived values for ion diffusion coefficients agree closely with other independent assessments.

Particle Diffusion (Real)

Even allowing for finite or infinite volume boundary conditions, real systems invariably involve the exchange of different counter-ions A

---

* erf = error function in a Gaussian distribution

and B. Hellferich was the first to allow for the fact that for non-isotopic ion exchange the different mobilities of the ions will give rise to a gradient of electric (charge) potential within the exchanger. This consideration together with the Fickian concentration gradient is essential for describing the coupling of ionic fluxes during an ion exchange reaction, and results in the general *Nernst–Plank* equation to describe particle diffusion kinetics, where grad ϕ is the electro-potential gradient:

$$\bar{J}_i = -\bar{D}_i \left(\text{grad } \bar{C}_i + \frac{z_i F \bar{C}_i}{RT} \text{grad } \phi \right) \quad (6.15)$$

The effect of this potential is to slow down the faster diffusing ion and accelerate the ion of lower mobility, thus balancing the net flux. Given that co-ions are absent from the resin phase and that the concentration of ionogenic groups is constant, equation 6.15 may be rewritten in terms of an interdiffusion coefficient $\bar{D}_{AB}$ such that:

$$\bar{J}_A = -\bar{D}_{AB} \text{ grad } \bar{C}_A \quad (6.16)$$

where

$$\bar{D}_{AB} = \frac{\bar{D}_A \bar{D}_B (z_A^2 \bar{C}_A + z_B^2 \bar{C}_B)}{z_A^2 \bar{C}_A \bar{D}_A + z_B^2 \bar{C}_B \bar{D}_B}$$

The interdiffusion constant $\bar{D}_{AB}$ is far from constant, its value depending greatly upon the ionic composition of the resin phase. For the infinite bath condition or trace exchange where $\bar{C}_B \ll \bar{C}_A$, the value of the interdiffusion coefficient equals that of the minority component $\bar{D}_B$ and *vice versa* for $\bar{C}_A \ll \bar{C}_B$. Thus the overall rate of exchange is governed by the diffusion coefficient of the ion of lowest concentration in the resin phase. Another experimentally proven prediction of the Nernst–Plank Theory is that the rate of exchange is faster when the initial ionic form of the resin is that of the ion of greatest mobility. Therefore, assuming $\bar{D}_A \gg \bar{D}_B$ for conversion of a resin in the A form, RA, to the B form, RB, the ratio of the forward and reverse rates of exchange is > 1 becoming larger as the exchange proceeds.

The flux condition given by equation 6.15 together with the constraints imposed by the mass balance, electroneutrality, and no net charge transfer gives a non-linear differential rate equation that is only amenable to computerized numerical solutions. Often sufficiently accurate kinetic data may be drawn from a 'linear driving force' model:

$$-\frac{d\bar{C}_i}{dt} = k_p(\bar{C}_i - \bar{C}_i^*) \quad (6.17)$$

or, the sometimes preferred 'quadratic driving force' approach proposed by Vermeulen:

$$-\frac{d\bar{C}_i}{dt} = k_p' \frac{(\bar{C}_i^2 - \bar{C}_i^{*2})}{2\bar{C}_i} \quad (6.18)$$

where k_p and k_p' are mass transfer parameters related to the particle geometry and the interdiffusion coefficient of the ions, whilst $\bar{C}_i^*$ represents the instantaneous equilibrium concentration of species i at the exchange surface.

For the infinite volume boundary condition and isotopic exchange, or trace exchange, ion selectivity does not influence particle exchange kinetics since the exchanger–solution interface is effectively devoid of the exchanger ion. However for macro-exchange ion selectivity grossly effects the phase boundary condition resulting in the preferred ion being more slowly released by the exchanger. In fact a consideration of selectivity can, by itself, have a great bearing upon the nature of the rate-controlling mechanism. For example, if the ion originally in the exchanger A is greatly preferred, the film side concentration of the entering ion B must increase greatly to provide a sufficiently high driving force for this ion to enter the resin. At the same time the build up of ion B at the interface reduces its concentration gradient across the film with respect to the bulk solution concentration of ion B. Therefore the loss of driving force across the film for the case $K_A^B \ll 1$, $\alpha_A^B \ll 1$ will encourage film diffusion rate control. For ion exchange between strongly functional resins and dilute solutions of strong electrolytes the effect of co-ion uptake upon particle diffusion kinetics is not greatly significant but greatly complicates the theory for concentrated solutions, exchange on weakly functional exchangers, and for co-ion participation in accompanying chemical reactions.

Film Diffusion (Ideal)

Interdiffusion within the film is treated as a quasi-stationary diffusion process across a planar layer (film). This means that the flux across the film adjusts itself rapidly to the changing boundary conditions and furthermore the flux is a function of the boundary conditions. Although qualitatively well understood, the theoretical treatment of film diffusion is made difficult by the fact that co-ions, as well as counter-ions, are present in the film. Furthermore, the selectivity, which is known to be a function of the degree of exchange, influences the counter-ion distribution at the exchanger–film interface.

From equation 6.1 and incorporating the constraints demanded by no net charge transfer, electroneutrality, and material mass balance the analytical rate equations for the infinite volume and finite volume boundary conditions are represented by equations 6.19 and 6.20 respectively:

$$\log(1 - F(t)) = -\frac{kt}{2.303} \quad (6.19)$$

and

$$\log(1 - F(t)) = -\frac{k't}{2.303} \quad (6.20)$$

where

$$k = \frac{3D}{r_o \delta}\left[\frac{C}{\bar{C}}\right]$$

and

$$k' = \frac{3D}{r_o \delta}\left[\frac{\bar{C}\bar{V} + CV}{\bar{C}V}\right]$$

The values of $F(t)$, r_o, δ, and t are as defined previously whilst D (m^2 s^{-1}) is the film diffusion coefficient, C (keq m^{-3}) the total external concentration, $\bar{C}$ (keq m$_r^{-3}$) the total concentration of exchanger functional groups, $\bar{V}$ (m^3) the volume of the exchanger, and V (m^3) the volume of the external solution. It is readily seen that the finite volume solution reduces to the infinite volume condition for $VC \gg \bar{V}\bar{C}$. Functions 6.19 and 6.20 are strictly only applicable to the conditions of no selectivity ($K_A^B = 1$, $\alpha_A^B = 1$) and constant diffusion coefficient ($D_A = D_B$), i.e. isotopic exchange. In either case a plot of $-\log(1 - F)$ against t will be linear of slope $k/2.303$ (s^{-1}) or $k'/2.303$ (s^{-1}).

Thus the analogy between formal diffusion theory, the 'linear driving force' model where the rate is proportional to $(C_A^* - C_A)$, and 'chemical kinetics' is very evident since the mathematical forms are the same. The difference arises in the interpretation of the gradient, namely, either in terms of a diffusion coefficient D (m^2 s^{-1}), a mass transfer coefficient D/δ (m s^{-1}), or a 'rate constant' k (s^{-1}) respectively.

Film Diffusion (Real)

For the exchange of ions of different mobilities the Nernst–Plank theory demands that account is taken of the electrical gradient established across the film. If the ion A originally in the exchanger RA is the faster ion a diffusion potential is established across the film, the net effect being to pull co-ions Y, and therefore electrolyte BY, out of the film hence lowering the concentration gradient of B in the film. This loss of driving force slows the forward rate of exchange. The opposite effect is observed for the reverse exchange which is the opposite behaviour to the case of particle diffusion control. For film diffusion the ion selectivity, since it effects the concentration of ions at the interface, also influences the rate of exchange. If ion A initially in the exchanger is greatly preferred ($K_A^B \ll 1$, $\alpha_A^B \ll 1$), then the liquid side concentration gradient of B is low which through a lack of driving force retards the overall rate of exchange. The opposite reasoning applies for a much greater preference by the exchanger for the ion initially in solution B.

The difference in ion mobilities, ion selectivity, and the effect of the diffusion potential upon co-ions as well as counter-ions make a rigorous theoretical treatment of film diffusion kinetics extremely difficult. Analytical rate equations have been proposed for monovalent ion exchange between ions of different mobility but no selectivity ($\alpha_A^B = 1$), and also for finite selectivity ($\alpha_A^B \neq 1$) but equal mobilities ($D_A = D_B$). A general solution for ion exchange between ions of different mobility and selectivity remains unavailable for either the finite or infinite volume condition. However, as Helfferich points out, the degree of accuracy attained by adopting the 'linear driving force' approach is sufficient for most purposes.

Although strictly an infinite volume solution for ions of equal mobility, a useful approximation for describing ion exchange under film diffusion control with selectivity is given by:

$$2.303 \log (1 - F) + \left(1 - \frac{1}{\alpha_B^A}\right) F = -kt \qquad (6.21)$$

where

$$k = \frac{3DC}{r_o \delta \bar{C} \alpha_B^A}$$

Figure 6.3 shows a plot of equation 6.21 for the system described in Table 6.1 from which the 'interdiffusion coefficient' is calculated to be 1.5×10^{-8} m^2 s^{-1}, compared with 4.0×10^{-9} m^2 s^{-1} found from equation 6.20. Obviously the calculated values for diffusion coefficients are affected by the choice of value for the film thickness δ, which may also be independently assessed by fluid dynamics theory. Usually the value for δ lies in the region of 10^{-5}–10^{-4} m.

In general, it can be seen that the effects of the diffusion potential and selectivity, whilst essential to a proper model for ion exchange, so complicate the theory that except for several distinctly defined boundary conditions, interpretation of real systems requires numerical analysis by means of a computer model. To advance the rate theory even further requires that one takes into account the effects of:

1. the variation of selectivity with ionic composition in either phase
2. the variation in the diffusivity characteristics of ions within the gel microstructure and macropores constituting macroporous exchangers

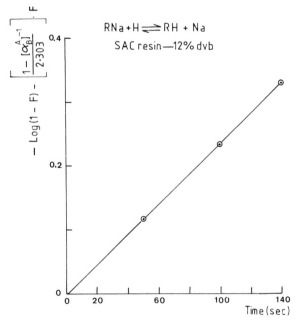

Figure 6.3 *Interpretation of ion exchange rate data according to film diffusion theory (equation 6.21) for sodium–hydrogen exchange on a styrenesulfonate resin (12% DVB)*
(Data from C. E. Harland, Ph.D. Thesis, University of Leeds, 1972)

3. solvent transfer during exchange through the effects of the diffusion potential and osmotic strength
4. electrolyte invasion of the exchanger
5. accompanying chemical reactions
6. ion–ion and ion–matrix interactions within the exchanger

Modern opinion views the Nernst–Plank theory as a special case of applying the thermodynamics of irreversible processes to ion exchange. It may also be argued theoretically and experimentally that the observed characteristics of ion exchange rate behaviour can only be fully explained by including chemical reaction as a flux-coupling mechanism as well as the diffusion potential. From a research standpoint it is most probable that future theoretical advances in ion exchange kinetics will result from the further application of non-equilibrium thermodynamics.

BOX 6.2 A Simple Rate Experiment

This experiment is used to demonstrate that ion exchange reactions are not instantaneous. They depend on the concentration and temperature of the solution used, and the diffusion of ions:

a) within the solution
b) across the Nernst or boundary film around the resin
c) within the resin bead

Under the conditions chosen here, the two latter effects are probably the controlling factors.

In the experiment:

$$R^-A^+ + B^+ \rightleftharpoons R^-B^+ + A^+$$

(where R^- represents the resin), the A^+ ions are released and are removed as they are produced. The rate of release of A^+ ions with respect to the time is recorded. The results obtained by this method will only demonstrate the principle in general terms and are not suitable for accurate determination of the reaction rates.

Requirements

Stirrer
Stop watch
Burette
Approx. 1 g of dry SAC resin (H form)
0.1 N sodium chloride

0.1 N sodium hydroxide
Indicator, methyl orange

Procedure

About 1 g of SAC resin is accurately weighed and transferred to a 250 ml beaker flask. 50 ml of deionized water, 5 ml of 0.1 N NaCl and methyl orange indicator are added. The stirrer and stop watch are started immediately. Titrations of the released acidity are carried out by adding excess 0.1 N NaOH and noting the time at which the methyl orange end point is reached. The procedure is carried on until no further acidity is released from the resin. A graph of titre versus time is plotted.

The amount of NaCl added at the beginning of the experiment may be varied and a number of equilibrium curves will be obtained.

TIME REQUIRED – 45 minutes

MECHANISM CRITERIA

Rate Equations

As evident from the preceding considerations the fit of experimental rate data to a particular rate theory equation does not necessarily prove the rate controlling mechanism. In the absence of previous knowledge of a system it is desirable to seek further mechanism qualifying data.

Reaction Half Times

The reaction *half time* $t_{1/2}$ is the time required to reach 50% attainment of equilibrium (F = 0.5). The half times for particle and film diffusion control are equal when both mechanisms proceed at the same rate in which case the ratio of the $t_{1/2}$ values will equal unity. Under infinite volume boundary conditions and counter-ions of equal mobility a dimensionless rate parameter is obtained from equations 6.12 and 6.21 given by:

$$\frac{\bar{C}\bar{D}\delta(5 + 2\alpha_B^A)}{CDr_o} = 1$$

The value of this quotient (Helfferich number) is $\gg 1$ and $\ll 1$ for film diffusion control and particle diffusion control respectively. For non-isotopic or non-trace exchange the values of $\bar{D}$ and D are best taken as the geometric mean of the self-diffusion coefficients in the exchanger and the film respectively. Strictly speaking, half time rate parameters should be derived for the specific boundary conditions which actually apply; hence some care is necessary in interpreting their value, especially when near unity, if the boundary conditions deviate greatly from theoretical.

Particle Size (r_0)

An examination of the rate theory equations for film and particle diffusion kinetics reveals an overall dependence of the rate of reaction on r_o^{-1} and r_o^{-2} respectively for gel ('homogeneous') exchangers. For a macroporous exchanger the rate of exchange, under particle diffusion control, may be either independent of particle size or vary as r_o^{-2} depending upon whether diffusion within the gel microsphere structure or macropores respectively is rate controlling.

Film Thickness (δ)

The rate of exchange under film diffusion control is inversely related to the thickness of the notional stagnant film surrounding the exchanger particles. If the stirring speed in a batch reaction, or superficial velocity within a packed column, is increased then the 'film thickness' is reduced with a consequent increase in exchange rate for a given system. A degree of caution should be taken when interpreting the effect of stirring rate since it is possible that a hydrodynamic limiting value for δ may be reached which is thereafter unaffected by further increases in the degree of agitation or stirring. Obviously particle diffusion rate control remains unaffected by the stirring rate.

Solution Concentration (C)

For film diffusion, but not particle diffusion, the rate of exchange is directly dependent upon the total external concentration. Consider a very small degree of exchange such that $F(t) \ll 1$; in which case the Maclaurin series expansion gives $\log(1 - F) \cong -F$, and the boundary conditions more nearly approach those for trace exchange (infinite volume). In such a case, equation 6.19 may be written:

$$F \cong \left(\frac{3DC}{2.303 r_o \delta \bar{C}}\right) t \qquad (6.22)$$

Hence, for a *fixed* small fractional resin conversion θ a plot of the initial rate of exchange as expressed by t_θ^{-1} against the external concentration C will be linear passing through the origin for film diffusion control. For particle diffusion kinetics the initial rate parameter will be independent of the external concentration. The absolute rates of exchange and the initial rate parameter $t_{0.08}^{-1}$ are shown in Figures 6.4 and 6.5 for the exchange between the sodium form of a 1% crosslinked (1% DVB) styrenesulfonate resin and hydrogen ions present as hydrochloric acid of different concentrations. The results clearly show the transition towards particle diffusion control with increasing external concentration. A similar approach was adopted by Reichenberg in 1952 for more highly crosslinked resins where solely

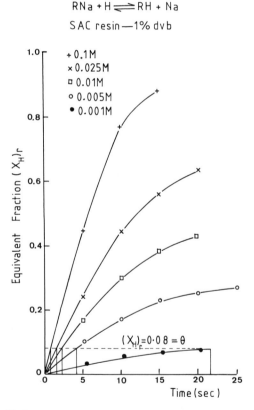

Figure 6.4 *Absolute rates of exchange for sodium–hydrogen exchange on a styrene-sulfonate resin (1% DVB) for different external concentrations of hydrogen ion*
(Data from C. E. Harland, Ph.D. Thesis, University of Leeds, 1972)

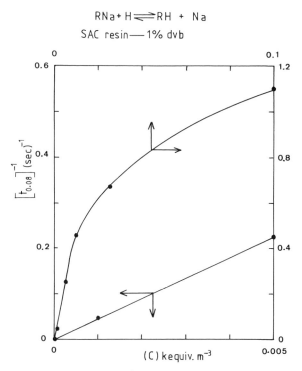

Figure 6.5 *Plot of initial rate t_θ^{-1} against external concentration C for sodium–hydrogen exchange on a styrenesulfonate resin (1% DVB), $\theta = 0.08$* (Data from C. E. Harland, Ph.D. Thesis, University of Leeds, 1972)

film diffusion control was shown to apply for external concentrations below approximately 0.02 keq m^{-3}.

Interruption Test

All the previously described mechanism criteria rely in some way or another upon comparing the predictions of rate 'theory' with the observations. The 'interruption test', first described by Kressman, is a favoured means of differentiating between film and particle diffusion rate control since it places no reliance upon theoretical boundary conditions or exchanger properties.

At a time whilst monitoring the exchange rate, remote from equilibrium, the exchanger and solution phases are temporarily separated and then re-equilibrated. If the rate of exchange is film diffusion controlled the pre-separation film concentration gradients are re-established almost instantly and the rate continues smoothly as if no interruption took place. However, if particle diffusion is rate controlling there is an initial increase in the rate of exchange (discontinuity), since during the period of interruption the particle concentration gradients (driving force) decay only to be restored over a finite short period following re-equilibration, as represented schematically in Figure 6.6.

COLUMN DYNAMICS

Most, but not all, commercial and laboratory applications of ion exchange involve operations where the solution to be treated (feed or influent) is passed downflow through a fixed bed (column) of ion exchange resin contained within a cylindrical vessel. The obvious difference between a stirred reactor (batch) contact of exchanger with the solution and a column process is that the former situation may be defined by a 'static' or true equilibrium. This is not the case with column operations since the position of equilibrium is always changing because:

1. The composition of a volume element of solution changes as it progresses down the column.
2. A given volume element of the column feed is always contacting unequilibrated resin as it passes down the column.
3. The displacement of the ionic products of the forward exchange

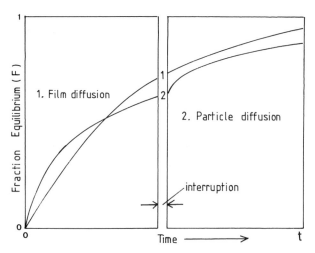

Figure 6.6 *The effect of interruption upon the rate of exchange according to controlling diffusion mechanism (schematic)*

reaction forces the exchange equilibrium to the right resulting eventually in complete conversion (exhaustion) of the resin commencing at the top of the column.
4. The extent of deviation from rapid attainment of local equilibrium within resin layers will be influenced greatly by the linear or specific flowrate (m h^{-1}) and the kinetics controlling the rate of exchange.

Consider a column of resin in the A form down which is passed an electrolyte solution BY such that ion exchange occurs between the counter-ions A and B, thus:

$$RA + BY \rightleftharpoons RB + AY \qquad (6.23)$$

AY and a decreased concentration of BY progress downwards to adjacent layers of exchanger whereby the concentration of BY is further reduced and the concentration of AY increases. At the top of the column the resin becomes progressively totally converted to the B ionic form owing to the continual displacement of AY.

The end result is as shown in Figure 6.7a where the top layers of resin are exhausted followed by an exchange zone across which the initial equivalent concentration C_o of ion B is reduced to zero whilst the concentration of ion A increases to C_o. Below the exchange zone, ideally, the resin is entirely in the A form and in equilibrium with the effluent AY, *i.e.* no exchange takes place. The sigmoid curve in Figure 6.7 representing the exchange zone is a schematic representation of the *effluent* concentration history, *not* the resin composition.

Eventually, the exchange zone is displaced to coincide with the column outlet such that any further passage of feed BY results in ion B appearing in the effluent at ever increasing concentration up to the limiting value of C_0, as shown in Figure 6.7b. If over the column history described above one were to plot the effluent concentration of ion B in keq m^{-3} against the volume throughput in m^3 a curve of the type shown in Figure 6.8 is obtained. Such a plot is termed a *breakthrough curve* for the system (RA + BY) under the given conditions.

Clayton proposes a graphical analysis of the breakthrough curve under favourable equilibrium conditions which states:

$$z = L\left(\frac{V_z}{V_b + fV_z}\right) \qquad (6.24)$$

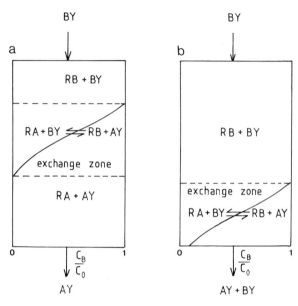

Figure 6.7 *Solution concentration profiles within an ion exchange column depicting the 'Ion Exchange Zone'*

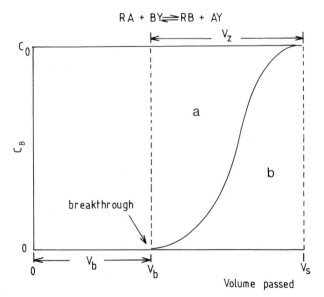

Figure 6.8 *A typical breakthrough curve (schematic)*

where

z = length of exchange zone (m)
L = column length (m)
V_b = volume of effluent to breakthrough of the ion being absorbed (m³)
V_s = total effluent volume to total loading of the resin bed (saturation) (m³)
$V_z = V_s - V_b$ (m³)

$$f = \frac{\text{area (a)}}{\text{area (a) + area (b)}} \qquad (6.25)$$

This treatment is similar to that derived by Michaels by a more rigorous theoretical appraisal. The area above the breakthrough curve up to the breakthrough point $V_b C_o$ is the resin loading of the absorbed ion B in keq at breakthrough. If the total resin (column) volume is V_R then C_{op}, the operating capacity of the column, is given by:

$$C_{op} = \frac{V_b C_o}{V_R} \text{ keq m}_R^{-3} \qquad (6.26)$$

This assumes of course that a column treatment is terminated at the first breakthrough of ion B which, in practice, is virtually always the case. In fact for largely practical and economic reasons some *leakage* of ion B is often present right from the start in which case 'breakthrough' is defined to be the point at which the level of B in the effluent becomes unacceptable. The notion of leakage or 'slip' is best left until the next chapter on the application of ion exchange and for now an idealistic view of breakthrough will suffice.

BREAKTHROUGH CURVE PROFILES

1. *Favourable Equilibrium:* If the ion exchange equilibrium is very favourable ($K_A^B \gg 1$, $\alpha_A^B \gg 1$) the exchange zone maintains a sharp profile as it advances down the bed. The breakthrough curve is little affected by the equilibrium and, all other considerations equal, is governed by the kinetics. Under conditions of constant column geometry and ion presentation rate, the profile of the exchange zone and therefore the breakthrough curve remains unchanged with time or column length and is said to show a *constant pattern* (Figure 6.9a). Under film diffusion control, being

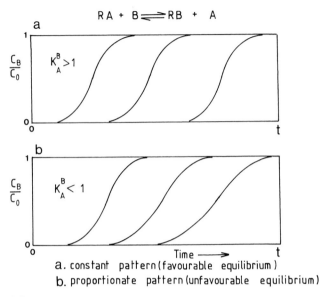

Figure 6.9 *Breakthrough curve profiles*

a. constant pattern (favourable equilibrium)
b. proportionate pattern (unfavourable equilibrium)

fast, the breakthrough capacity is reasonably insensitive towards flowrate for favourable equilibria on strongly functional resins. The opposite case often applies to exchange on weakly functional resins where slower particle diffusion is rate controlling and breakthrough capacity (operating capacity) decreases with increasing ion presentation rate. In other words the performance of such a resin is said to be rate sensitive.

2. *Unfavourable Equilibrium:* ($K_A^B \ll 1$, $\alpha_A^B \ll 1$). In this case the exchange front or boundary does not remain sharply defined as it progresses down the column. Now equilibrium properties, rather than kinetics, dominate in that ion B is able to overtake ion A and the exchange zone for complete removal of ion B is lengthened and not self-sharpening. The breakthrough curve does not demonstrate a constant profile but is broadened with increasing residence time for a given presentation rate and is said to show a *proportionate pattern* (Figure 6.9b). This type of behaviour often prevails whilst regenerating a column with a concentrated solution of the least preferred ion although the 'electroselectivity' effect for heterovalent ion exchange can greatly influence the equilibrium.

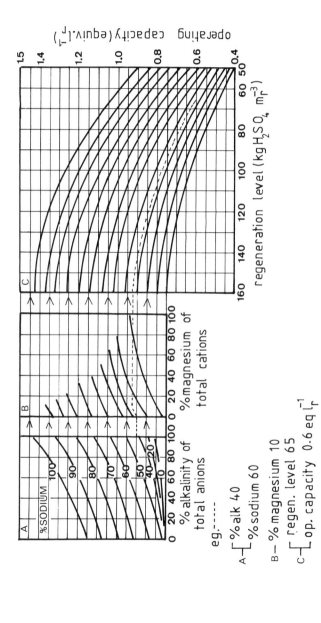

Figure 6.10 A linear nomography representation of coflow column operating capacity for cation exchange (Ca, Mg, Na) against a standard form styrene sulfonic acid resin (8% DVB) as applied to water demineralization (By permission of The Permutit Company Ltd)

PROCESS DESIGN

Clearly the shape and operating capacity as defined by the breakthrough curve is very dependent upon such factors as:

a) bed-depth
b) exchanger characteristics
c) flowrate
d) kinetics
e) selectivity towards influent ions
f) initial resin composition

The study of how the considerations listed above affect column performance denotes a departure from the 'pure science' of ion exchange towards the *technology* of the subject and is the domain of the chemical engineer. For fixed bed ion exchange in water treatment applications the data for obtaining column operating capacity to a desired 'breakthrough' is usually presented as a sequence of graphs and empirical 'correction factors' or nomogram relationships as illustrated by Figure 6.10 for hydrogen ion exchange. This and a copious quantity of similar data for water treatment ion exchange have arisen from exhaustive column performance assessments by resin manufacturers and engineering companies and apply to equipment designs which have been established over some fifty years or more.

In other areas of technology such as chromatographic separations, and hydrometallurgy which may often involve resin transfer (movement) operations, fluidized beds, and truly countercurrent contact of resin and liquid, the acquisition of design data is more often through a fluid dynamic and mass transfer approach to describe, and thereby scale up, column designs in order to realize commercial practices. A text of this nature is neither the vehicle for, nor can it do justice to, such an important topic and the reader is referred to the bibliography for a more detailed account of column dynamics.

FURTHER READING

F. Helfferich, 'Ion Exchange', McGraw–Hill, London and New York, 1962.

F. G. Helfferich and Yng-Long Hwang, 'Ion Exchange Kinetics', in 'Ion Exchangers', ed. K. Dorfner, Walter de Gruyter, Berlin and New York, 1991, Ch. 6.2, p. 1277.

F. Helfferich, 'Ion Exchange Kinetics', in 'Ion Exchange (A Series of Advances)', ed. J. A. Marinsky, Edward Arnold (London), Marcel Dekker (New York), Vol. 1, 1966, p. 65.

M. J. Slater, 'Principles of Ion Exchange Technology', Butterworth–Heinemann, Oxford, 1991.

A. S. Michaels, 'Simplified Method of Interpreting Kinetic Data in Fixed-Bed Ion Exchange', *Ind. Eng. Chem.*, 1952, **44**, p. 1922.

M. Streat, 'General Ion-Exchange Technology', in 'Ion Exchangers', ed. K. Dorfner, Walter de Gruyter, Berlin and New York, 1991, p. 685.

R. R. Harries, 'Ion Exchange Kinetics In Condensate Purification', *Chem. Ind. (London)*, 1987, No. 4, 104.

Chapter 7

Some Basic Principles of Industrial Practice

INTRODUCTION

The remaining chapters aim to explain how and where ion exchange is applied in the process industries. Rather than attempting an exhaustive survey of all documented processes it is proposed to explain, with the support of selected applications, the underlying process principles with particular emphasis placed upon the influence of exchanger characteristics, equilibria, and kinetics. Equipment and engineering criteria are left to be discussed under Chapter 10 and for the present, unless otherwise stated, it may be regarded that contact between the ion exchanger and the solution to be treated occurs across fixed beds of ion exchange resin contained in columns through which the solution is passed.

Ion exchange processes can be conveniently classified under four general headings:

1. *Purification and Ion Removal.* Commonly purification processes involve removal of undesirable ions by exchange for an acceptable ion which appears in the column effluent. Water treatment and many waste treatment processes come into this broad category.
2. *Recovery.* Contrary to 1. above, in recovery processes the desired product is more often than not retained on the exchanger to be subsequently displaced or eluted by another suitable ion. The most familiar examples of such processes are associated with the field of hydrometallurgy and the refining of pharmaceutical agents.
3. *Metathesis.* Metathesis is rather a general case and is simply the replacement of one ion (cation or anion) of a single electrolyte by another by means of ion substitution using an exchanger in the appropriate ionic form. Many preparative and analytical chemistry

techniques, as well as some industrial practices, employ this simple expedient.

4. *Separation.* Separation by ion exchange is achieved by firstly allowing the exchanger to take up a quantity of mixed ions which is far less than the total exchange capacity of the exchanger. The mixture is then resolved into separate zones within the column by continuous displacement or elution with a solution of another ion. In this process known as *ion exchange chromatography* each ionic component of the mixture appears in the effluent in order of increasing affinity for the resin concerned.

COLUMN OPERATIONS

Before proceeding to outline specific processes it is as well to explain some terms which commonly arise when describing ion exchange unit operations. A sound basis is afforded by the column process represented in general terms by Figure 7.1 where, by way of example, it is required to remove cations B^+, C^+, and D^+ from an aqueous solution of mixed electrolytes BX, CY, and DZ by exchange for cation A^+ on the resin. For the sake of simplicity the cations are assumed monovalent and in association with co-ions X^-, Y^- and Z^-. Clearly an analogous anion exchange example would be equally valid, as would be the more general case of heterovalent exchange.

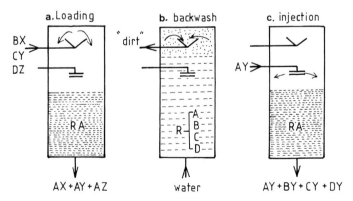

Figure 7.1 *Column Operations:*
  a) Loading
  b) Backwashing
  c) Chemical injection

Loading (Figure 7.1a)

Operations commence with passing the mixed electrolyte solution down the column of ion exchange resin whereby exchange occurs giving a continuous flow of effluent of mixed composition AX, AY and AZ of *total equivalent concentration* (keq m^{-3}) equal to that of the column *feed* or *influent*. The column effluent may either by the final product or may become the feed for further ion exchange processing and therefore this operation is often termed the *service cycle*. At some later time the operating capacity of the column is reached indicated either by a preset volume treated (*metered capacity*) or by a deterioration of product quality to below predetermined limits (*breakthrough capacity*). At this point the run is stopped and the column taken out of service and made ready for *regeneration*. Of course, the loading process may be continued by switching to another column or *standby plant*, or as is sometimes the case, the equipment is of such a design to allow a truly continuous cycle.

At termination of the loading cycle under fixed bed operation the column remains flooded with partially treated feed. In processes where the feed liquor commands a high monetary value with respect to its constituents, as in for example, such processes as carbohydrate refining, metal recovery, and pharmaceutical production it is common practice to '*sweeten off*' the column by displacing up to about one bed volume of column residual which is saved for reprocessing.

Backwashing (Figure 7.1b)

With some exceptions mentioned later, the first stage in the regeneration sequence of operations is *backwashing* with a controlled upflow of water to give typically 30–40% bed expansion. Backwashing a bed, where applicable, serves to effect the following important requirements:

1. To flush out particulate matter entering with the feed which has been filtered out across the top few centimetres of the resin bed. An ion exchange bed is not a filter in the accepted unit operation sense and therefore can only tolerate about 2–3 g m^{-3} of suspended solids. Plugging of resin beds through excessive dirt loading gives rise to an unacceptable differential pressure across the bed (*pressure drop*) which not only can cause a reduction in flow but may generate a sufficiently high net compressive force so as to break resin beads and maybe even distort or break internal fabricated components. Pressure drop considerations will influ-

ence resin selection as discussed previously in Chapter 4, but in any event it is not desirable to operate at differential pressures across the resin bed alone much in excess of 170 kPa (1.7 bar).
2. To remove small broken fragments of ion exchange resin or '*resin fines*' arising through normal anticipated resin degradation, but only fragments small enough to be carried out with the upward flow will be removed.
3. To grade the bed whereby broken resin which is too coarse to be washed out completely accumulates at the bed surface from where it may be removed by scraping. Regular '*bed dressing*' to remove broken resin and 'fines' is extremely important since their unchecked build up within the column will cause increased pressure drop which accelerates further resin breakage.
4. To decompact, and relax the resin bed thereby reducing pressure drop and eliminating any tendency towards encouraging preferential flow paths, or *channelling*, which would otherwise lead to a reduced utilization of resin exchange capacity and premature column breakthrough.
5. Finally, backwashing is employed to grade resin beds in cases where the process design calls for maintaining a separation between different types of resin within the same column.

Chemical Injection (Figure 7.1c)

Chemical injection is the next step of the regeneration cycle during which an appropriate electrolyte solution of concentration typically between 5–150 kg m^{-3} is passed through the resin to convert it to its initial ionic form RA ready for the next treatment cycle. Prior to commencing the next run the column is given a short *regenerant displacement rinse* followed by a longer *fast rinse* to purge the resin bed of last traces of regenerant chemical. Just as feed may be conserved by 'sweetening off' a column prior to regeneration so '*sweetening on*' may apply at the beginning of the loading cycle when the first one or two bed volumes of the column product are discarded or recycled since this early product is greatly diluted by displaced rinse effluent (commonly water).

OPERATING CAPACITY AND REGENERATION EFFICIENCY

With some exceptions, the loading or service cycle in most industrial processes involve favourable ion exchange equilibria. It follows that the reverse exchange, or regeneration, is unfavourable. For univalent–univalent systems a satisfactory degree of regeneration (resin conversion) is achieved through a mass action effect brought about using a relatively high concentration of regenerant chemical. In dilute heterovalent systems the extremely high preference shown by strongly functional resins towards the ion of greatest charge would suggest that regeneration with a monovalent ion is impractical. However, the '*electroselectivity*' effect discussed in Chapter 5 ('Selectivity Coefficient') shows the separation factor expressed for ion exchange in the direction of uptake of the polyvalent ion to be inversely related to the total solution concentration (equation 5.20) thus making regeneration with a sufficiently concentrated monovalent regenerant far easier. Therefore the 'electroselectivity' effect may be employed to advantage in both the loading and regeneration cycles, and this is a principal reason why ion exchange proves to be so successful in water treatment operations. The above considerations apply essentially to strongly functional resins rather that weakly functional or chelating resins, whose equilibrium exchange properties are often governed more by 'unique' selectivity behaviour.

A practical consequence of strongly functional resin equilibrium characteristics is that, in commercial operations, the degree of regeneration (resin conversion) never equals the total wet volume capacity available. To achieve this would require so much excess chemical above the equivalent stoichiometric requirement as to make most processes uneconomic. The greater the amount of useful exchange capacity (operating capacity) obtained from passing a given quantity of regenerant chemical the more efficient the regeneration, and the greater is the volume throughput per loading or service cycle. Therefore *regeneration efficiency* may be defined as:

$$\text{Regeneration Efficiency} = 100 \times \frac{\text{Regenerant consumed (keq)}}{\text{Total regenerant passed (keq)}}$$

or

$$\text{Regeneration Efficiency} = 100 \times \frac{\text{Regenerant consumed (keq m}^{-3}{}_R)}{\text{Total regenerant passed (keq m}^{-3}{}_R)}$$

The ratio of the total quantity of regenerant chemical passed (keq) to the volume of the resin column ($m^3{}_R$) is known as the *regeneration level* ($keq\, m^{-3}{}_R$). Obviously the quantity of regenerant consumed per unit volume of resin ($keq\, m^{-3}{}_R$) is the degree of resin conversion during regeneration and is equivalent to the operating capacity achievable over the next column loading cycle for a given influent analysis.

Figure 7.2 shows typical column operating capacity curves for the exchange of sodium, magnesium, and calcium ions, all as chlorides, against the hydrogen form of a gel styrenesulfonic acid resin (8% DVB) for various regeneration levels using 0.15 Molar sulfuric acid. Several fundamental features are immediately noticeable. Firstly, at very low regeneration levels the exchanger is able to exchange most of the hydrogen ions with relatively little breakthrough of unutilized acid, but of course, the degree of resin conversion, and therefore the resulting column operating capacity is low.

At higher, more practical regeneration levels the extent of resin conversion will be greater but at the expense of greater acid breakthrough or reduced regeneration efficiency. For the examples cited, the regeneration efficiencies (E) are given by the gradients of chords (OA, OA′, OA″) and (OB, OB′, OB″) at regeneration levels of 40 kg(H_2SO_4) m^{-3}_R and 120 kg(H_2SO_4) m^{-3}_R respectively (see Figure 7.2). A different situation arises with weakly functional cation and anion exchange resins which exhibit very favourable regeneration

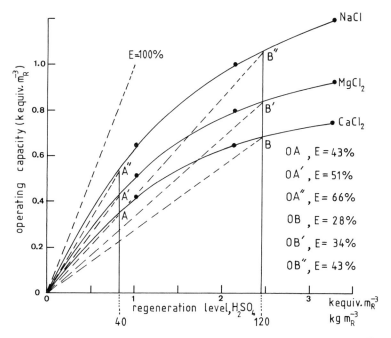

Figure 7.2 *Operating capacity curves for neutral cation exchange on a styrenesulfonic acid resin (8% DVB) for different regeneration levels of sulfuric acid*

equilibria with acids and alkalis respectively and therefore offer high regeneration efficiencies.

Notice also that for a given regeneration level the operating capacity achieved is least for divalent calcium compared with monovalent sodium which reflects, independent of the electroselectivity effect, the greater affinity of the resin for polyvalent cations. The same general considerations also apply to the heterovalent exchange between simple anions on strongly functional resins. The influent ionic composition during the loading cycle with respect to counter-ions *and co-ions* does influence the operating capacity, but for any *given* feed analysis the performance of industrial plant is invariably *regeneration controlled*. Thus a 'trade off' has always to be made between column operating capacity and the cost incurred in regenerant chemicals.

Basic Economics

Suppose for any given process the operating capacity is Q (keq m$^{-3}_R$) and the bed volume is R (m^3); then the gross treated volume, V (m^3), during the loading or service cycle will be given by:

$$V = \frac{RQ}{L} \qquad (7.1)$$

where L (keq m^{-3}) is the total ion concentration in the feed. Therefore every volume V (m^3) treated would cost a sum of money in chemicals to regenerate or elute the column, and this consideration immediately demonstrates the value of achieving the optimum regeneration efficiency conducive to the purity of product required. It should be equally obvious that the greater the total concentration of exchangeable ions in the feed the smaller will be the volume treated per loading cycle hence incurring frequent regenerations and therefore high chemical costs per unit product. A not so immediately obvious factor is that the time taken to regenerate a column can sometimes take several hours during which time it is off-line. Therefore high feed concentrations and the need for frequent lengthy regenerations combine to drastically reduce the net output from a given ion exchange unit.

In as much as a high influent ion burden greatly increases both the capital cost of plant (large columns and large resin inventory) and operating costs (frequent regenerations) commercially viable ion exchange processes are therefore confined to the treatment of dilute solutions, say less than 40 eq m^{-3} and more often than not less than

10 eq m^{-3}. Therefore for any process, the size, and therefore the cost, of a plant will be determined principally by such factors as:

1. total concentration of exchangeable ions
2. net continuous treatment rate required
3. ion exchange equilibria concerned
4. regeneration efficiency obtainable
5. ion exchange resin selection

COLUMN BREAKTHROUGH AND 'LEAKAGE'

Loading Cycle

Consider again the ideal case of column ion exchange as depicted by Figure 7.1a, and further assume that the apparent selectivity coefficient values K for the system are such that the affinities of ions B^+, C^+, and D^+ for the resin in its regenerated RA form are such that, $K_A^D > K_A^C > K_A^B > 1$. As loading takes place the preferred exchange of influent ions for ion A^+ will be in the order $D^+ > C^+ > B^+$, and consequently the exchange zone length (z) required for complete exchange of influent ions will increase in the order $z_B > z_C > z_D$. Furthermore, any given influent ion will displace another of lower affinity for the resin.

The overall result is that at column breakthrough ('exhaustion') a qualitative description of the column is as illustrated, purely schematically, in Figure 7.3 where the hatched area represents exhausted resin

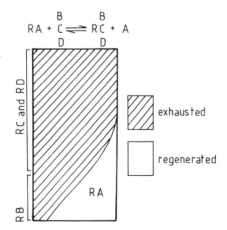

Figure 7.3 *Distribution profiles (schematic) of exhausted and regenerated resin at column breakthrough*

and the plain area resin remaining in its regenerated form. Exhausted resin now present at the outlet end of the column means that any further exchange will be incomplete and the ion of lowest affinity, B^+, will be the first to appear in the effluent signifying breakthrough. In other words, it can be seen that the overall resin condition at breakthrough is such that the column outlet region of the resin is enriched in the ion of lowest affinity and conversely at the inlet end of the column. Theoretical breakthrough may be defined as the point during the loading cycle where the column is unable to accommodate the exchange zone length required to effect complete removal of the influent ions.

Regeneration Cycle (Coflow)

Coflow is the term given to the mode of regeneration where the direction of flow of the regenerant chemical *and the rinse* is the same as for the loading or service cycle (usually downflow). During coflow regeneration the fresh concentrated regenerant firstly contacts the top of the resin bed which is most heavily exhausted with ions of greatest affinity for the resin. Although the regeneration equilibria become less unfavourable as the regenerant passes down the bed its concentration is increasingly depleted allowing back exchange to occur. Reverse exchange also favours the ions of highest affinity for the resin (C^+ and D^+) and therefore ion B^+ is both eluted by the regenerant ion A^+ and displaced by ions C^+ and D^+.

Finally the competing back reaction is assisted by the fact that ions eluted or displaced from the upper regions of the bed have to traverse the entire column length before emerging at the outlet. The overall effect of coflow regeneration is such that upon completion of regeneration the top of the bed is reasonably highly converted but the opposite holds at the column outlet where the resin is only partially regenerated and contains a significant fraction of B^+ ions.

The consequences of the coflow regeneration pattern become immediately apparent at the start of the next loading cycle. At the top of the column ion exchange is complete giving an equivalent concentration of A^+ ions which then travel unaffected down the bed:

$$\text{RA} + (B^+, C^+, D^+) \begin{pmatrix} X^- \\ Y^- \\ Z^- \end{pmatrix} \rightarrow R \begin{pmatrix} B \\ C \\ D \end{pmatrix} + A^+ \begin{pmatrix} X^- \\ Y^- \\ Z^- \end{pmatrix} \quad (7.2)$$

Upon nearing the outlet, ions A^+ encounter resin partially in the B^+ ionic form and ion exchange can occur allowing a quantity of ion B^+ to appear in the column effluent (product):

$$RB + A^+ \begin{pmatrix} X^- \\ Y^- \\ Z^- \end{pmatrix} \rightarrow RA + B^+ \begin{pmatrix} X^- \\ Y^- \\ Z^- \end{pmatrix} \qquad (7.3)$$

This mechanism by which the complete exchange for ion A^+ is spoilt by reverse exchange across a layer of insufficiently regenerated resin at the bottom of the column is termed *elution leakage*, or '*slip*'. It is important to grasp that the phenomenon of elution leakage is due to the presence of partially regenerated resin near the column outlet and not incomplete exchange during loading. The problem of elution leakage arises because for acceptable regeneration levels there will always be a fraction of resin at the bottom of the bed which after the regeneration cycle remains unregenerated and partially loaded with influent ions of *lowest* affinity for the resin. For film diffusion controlled kinetics, elution leakage may be expected to be reasonably independent of flowrate at a given instant in the cycle.

The column history during loading and regeneration is shown schematically in Figure 7.4a where initially, leakage reduces as ion B^+ is eluted from the bottom of the bed but, eventually, as loading proceeds the 'exhaustion' profile reaches the column outlet and the presence of ion B^+ in the effluent increases again. Therefore if the effluent quality is monitored the initial leakage and subsequent exhaustion profile would take the form shown in Figure 7.5a. The terms elution and displacement are often confused and for many processes it is important to discern between their correct meanings. *Displacement* occurs when an ion on the resin is exchanged for another of greater affinity (selectivity) for the exchanger, whereas *elution* is exchange for an ion of lower affinity.

Finally, it is possible for 'leakage' to occur because of incomplete ion exchange across a regenerated region of the bed in which case it is termed *kinetic leakage*. This commonly arises if the ion exchange resin is fouled, or the flowrate is too great especially if particle diffusion kinetics are rate controlling. In either case 'leakage' is flow dependent decreasing at reduced flowrates. Kinetic leakage may even result in the most favourably exchanged ion appearing prematurely in the effluent.

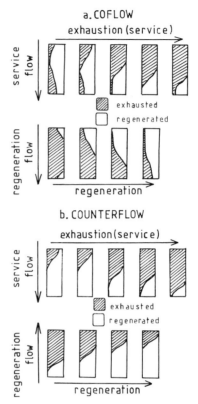

Figure 7.4 *Resin loading distribution profiles (schematic) during loading and regeneration*
 a) *Coflow mode*
 b) *Counterflow mode*

Regeneration Cycle (Counterflow)

Counterflow is the term ascribed to the mode of regeneration where the direction of flow of the regenerant chemical *and the rinse* is opposite to that for the loading or service cycle. Assuming the loading cycle to be downflow (but upflow service–downflow regeneration is equally viable) the regenerant is passed upwards from the service outlet end of the column. Fresh, concentrated regenerant now first comes into contact with that region of the resin column which is least exhausted and only by ions of *least affinity* for the resin. Therefore the overall regeneration efficiency, and the degree of conversion across the outlet region of the resin is high. Furthermore the chromatographic banding

Some Basic Principles of Industrial Practice

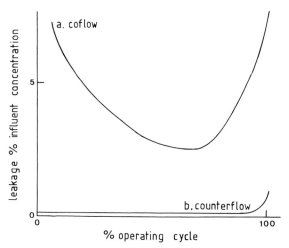

Figure 7.5 *Typical leakage profiles (schematic) during the loading cycle*
  a) Coflow mode
  b) Counterflow mode

within the column occurring over the loading cycle assists regeneration in the counterflow mode. Also ions of greatest affinity eluted from the upper regions of the bed have a shorter distance to travel, and therefore less time to back exchange, before passing out of the column to drain.

The column history under counterflow regeneration is illustrated by Figure 7.4b which now shows the service outlet zone of the bed to be highly regenerated and consequently the elution leakage over the following loading cycle is greatly reduced as shown by Figure 7.5b. Therefore the advantages offered by counterflow compared with coflow regeneration are:

1. much improved regeneration efficiency
2. lower resin inventory
3. greatly improved treated product quality

In order to sustain the benefits from fixed bed counterflow regeneration it is imperative that the resin bed remains consolidated and undisturbed at all times during loading and reverse flow regeneration. Some ways this is achieved are discussed later in Chapter 10, but it follows that conventional backwashing of a counterflow designed column is undesirable since the inevitable bed disturbance causes 'exhausted' resin to mix with the lightly loaded resin zone at

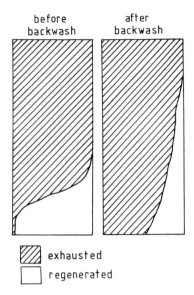

Figure 7.6 *The effect of backwashing upon a counterflow regenerated column profile*

the bottom of the bed as depicted in Figure 7.6. Should it be necessary to carry out a full bed backwash of a counterflow designed bed, then a following regeneration with at least twice the normal quantity of regenerant chemical has to be carried out in order to ensure low leakage characteristics over the next service cycle. Finally, in order to maintain virtually 100% regeneration of resin across the service outlet region of the column particular attention has to be paid to the purity of the regenerants and rinse waters used so as not to introduce undesirable ions.

Chapter 8

Water Treatment

The best known and most widely encountered application of ion exchange is in the field of Water Treatment. It has been estimated that the Worldwide production of ion exchange resins is of the order of 500 000 m^3 per annum of which probably in excess of 90% is employed in such water treatment processes shown collectively in Table 8.1. The majority of applications are for 'clean' or 'natural' water treatment as opposed to waste water or effluents which are waters made 'dirty' by the inclusion of added chemicals and solids arising from manufacturing processes. Of course, natural waters are

Table 8.1 *Ion exchange processes in water treatment*

Process	Resin type
Softening	Strong Acid Cation* Weak Acid Cation Chelating
Dealkalizing	Weak Acid Cation* Strong Base Anion
Demineralizing	Strong Acid Cation* Weak Acid Cation Strong Base Anion* Weak Base Anion
Organic removal	Strong Base Anion* Weak Base Anion
Nitrate removal	Strong Base Anion
Oxygen removal	Strong Base Anion

* Principal resin type used

not pure but contain suspended solids, dissolved gases, colloidal matter, and dissolved minerals, or salts, to an extent governed largely by the nature of its source. Raw water supplied to industry in the UK arises from one or more of several principal sources:

1. *Natural catchments* of rainfall and land surface water such as lakes and rivers.
2. *Man-made catchments*, e.g. impounding reservoirs.
3. *Subterranean sources* e.g. wells and boreholes.
4. *Municipal supplies* whilst often arising from one or several of the sources listed above are frequently pretreated by the Regional Water Authority in order to meet the requirements for potable quality laid down by The European Community directives given in Table 8.2.

WATER ANALYSIS

The physical characteristics of a given water supply such as turbidity, colour, and suspended solids are extremely important considerations since they determine whether or not some kind of pretreatment is required prior to ion exchange, for example coagulation and filtration. The chemical characteristics are given principally by the mineralogical analysis which quantifies the concentration of ions present, and impacts directly upon the size and design of ion exchange plant as well as identifying potential problem areas arising from the presence of resin foulants. In some critical applications it is necessary to carry out a bacteriological and biological analysis of an intended supply in order to assess requirements for ancillary treatments and to guard against possible bacteriological fouling of resin beds and items of plant. Water chemistry is a separate subject in its own right and too extensive to enter into here except to briefly explain some analytical terms in order that someone new to ion exchange, as applied to water treatment, might understand both the jargon and some of the strange units which still persist in the literature.

A water analysis will usually report the levels of ionic constituents in terms of their actual mass concentration by volume in milligrams per litre $(mg\,l^{-1})$ or parts per million (ppm). Very low concentrations of trace constituents are often reported as micrograms per litre $(\mu g\,l^{-1})$ or parts per billion (ppb). Numerically the values for $(mg\,l^{-1}$ and ppm) or $(\mu g\,l^{-1}$ and ppb) are taken as being equal for very dilute aqueous solutions of density close to that of pure water at 4 °C, although by definition the units are not identical.

Table 8.2 *EC Directives for potable water quality*

Determinand	Permitted concentration value (entering distribution)	
pH	5.5–9.5	pH units
Conductivity	1500	$\mu s\,cm^{-1}$
Turbidity	4	FTU
Colour	20	Hazen
Nitrogen Ammonia	0.5	$mg\,l^{-1}\,NH_4^+$
Nitrate	50	$mg\,l^{-1}\,NO_3^-$
Nitrite	0.1	$mg\,l^{-1}\,NO_2^-$
Chloride	400	$mg\,l^{-1}\,Cl$
Fluoride	1.5	$mg\,l^{-1}\,F$
Sulfate	250	$mg\,l^{-1}\,SO_4^{2-}$
Sodium	150	$mg\,l^{-1}\,Na$
Potassium	12	$mg\,l^{-1}\,K$
Copper	3	$mg\,l^{-1}\,Cu$
Magnesium	50	$mg\,l^{-1}\,Mg$
Calcium	250	$mg\,l^{-1}\,Ca$
Zinc	5	$mg\,l^{-1}\,Zn$
Cadmium	0.005	$mg\,l^{-1}\,Cd$
Aluminium	0.2	$mg\,l^{-1}\,Al$
Lead (Unflushed)	0.1	$mg\,l^{-1}\,Pb$
Lead (Flushed)	0.05	$mg\,l^{-1}\,Pb$
Chromium	0.05	$mg\,l^{-1}\,Cr$
Manganese	0.05	$mg\,l^{-1}\,Mn$
Iron	0.2	$mg\,l^{-1}\,Fe$
Nickel	0.05	$mg\,l^{-1}\,Ni$
Coliforms	0	No. $(100\,ml)^{-1}$
E. Coli	0	No. $(100\,ml)^{-1}$
Faecal *Streptococci*	0	No. $(100\,ml)^{-1}$
Trihalomethanes (THM)	0.1	$mg\,l^{-1}$
Polycyclic Aromatic Hydrocarbons (PAH)	0.0002	$mg\,l^{-1}$
Phenols	0.0005	$mg\,l^{-1}$
Pesticides	0.0005	$mg\,l^{-1}$

In terms of calculating the total amounts of different ions that a resin can exchange over a given cycle, absolute concentrations are of little use since they are incompatible and cannot be summated to give a total loading. An analogy may be drawn with monetary currencies whereby, upon returning from a European touring holiday, it is impossible to empty one's pockets of say, loose Swiss francs, Austrian schillings, and Spanish pesetas, and immediately assess their total value – not unless firstly conversions are made to a common currency. Therefore, ion concentration must be converted to the 'common currency' of chemical equivalents. The equivalent mass of an ion is equal to the ion atomic mass divided by its electrical charge *as exchanged by the resin*, from which equivalent concentrations in milliequivalents per litre ($meq\,l^{-1}$) are found from the simple relationship:

$$\frac{\text{ion concentration (mg}\,l^{-1})}{\text{ion equivalent mass}} = \text{ion concentration (meq}\,l^{-1})$$

where

$$\text{ion equivalent mass} = \frac{\text{ion atomic mass}}{\text{ion charge}}$$

The ($CaCO_3$) Convention

The scientist would almost certainly regard equivalent concentrations to be logical and acceptable, but water treatment had its beginnings as much in engineering as chemistry and several different units arose which unfortunately are still commonplace. The most common of these is concentration expressed as milligrams (calcium carbonate) per litre or parts (calcium carbonate) per million; written mg ($CaCO_3$) l^{-1} and ppm($CaCO_3$) respectively. The reasons for this are obscure, but the traditional process of water softening by the addition of calcium hydroxide (hydrated lime) produces a precipitate of calcium carbonate by reaction with calcium hydrogencarbonate in the water:

$$Ca(HCO_3)_2 + Ca(OH)_2 \rightarrow 2\,CaCO_3(s) + 2\,H_2O \qquad (8.1)$$

In order to calculate the 'lime' dose required and the quantities of 'sludge' produced for a given water the practice emerged of expressing the stoichiometry of the chemical reaction in terms of ppm($CaCO_3$) – another form of 'common currency'. The molecular mass of calcium carbonate is 100 which might also have been an influence upon adopting this practice. The equivalent mass of a chemical compound, as opposed to a discrete ion, is the sum of the equivalent masses of its constituent ions which for $CaCO_3$ is 50, given by:

$$Ca^{2+} + CO_3^{2-} = CaCO_3$$

equiv. masses $\qquad \dfrac{40}{2} + \dfrac{60}{2} = 50 \qquad (8.2)$

Therefore a conversion from ionic concentration (ppm) to

ppm($CaCO_3$), or mg($CaCO_3$) l^{-1} is given by:

$$\text{Ion Conc. (ppm)} \times \frac{50}{\text{ion equivalent mass}} = \text{Conc. (ppm} CaCO_3)$$

(8.3)

Thus, when confronted with a water analysis it is necessary to decide upon a consistent convention to express ion concentrations in order to calculate resin loadings and plant capacities; and it follows that the same convention must be used to express ion exchange resin operating capacities. Industrial practices in general adopt either the 'chemical equivalent' or '$CaCO_3$' conventions for expressing solution and resin compositions. However, in water treatment especially, the position is made even more confused by a dogged survival of other derived units based on pounds mass (lb), cubic feet (ft^3), gallons (gall.), grains (gr.), and kilograins (kgr.), which the author will not encourage by discussing further except to draw attention to conversion tables in the Appendix.

Table 8.3 shows the principal cationic and anionic species occurring in all natural waters to a greater or lesser extent together with their concentrations for an actual supply in $mg\,l^{-1}$, $meq\,l^{-1}$, and ppm($CaCO_3$). The principal components may be categorized according to the following conventions which are widely adopted in the context of water treatment by ion exchange. The main ionic constituents are alkaline (hydrogencarbonate, carbonate) and neutral (chloride, sulfate, and nitrate) salts of the metals calcium, magnesium, sodium, and sometimes potassium. The total concentration of calcium and magnesium ions is termed the *total hardness, TH*.

The alkaline salts (hydrogencarbonates, and possibly carbonates) of calcium and magnesium are titratable against a standard mineral acid using traditionally methyl orange indicator giving the *total alkalinity* or (M) alkalinity, *M.ALK*. The total alkalinity is sometimes called *temporary hardness* as it may be deposited from the water as precipitated calcium carbonate by the action of heat. Normally, the total hardness is greater than the total alkalinity, the difference being the amount of neutral (non-alkaline) hardness salts present often called *permanent hardness*. Should the total alkalinity exceed the total hardness then only alkaline hardness is present, the excess alkalinity being due to alkaline sodium salts (hydrogencarbonate, carbonate, and even possibly hydroxide). The alkaline salts present are in buffered equilibrium with dissolved carbon dioxide and this determines the pH of the water which is usually in the range 6.5–8.5.

Consider the result of passing a sample of raw water through a

Table 8.3 Principal ionic species in natural UK waters together with concentration values for a specific UK municipal supply ($pH = 7.9$)

Group of ions	Cations				Group of ions	Anions			
	Ion	mg l^{-1} (ppm)	meq l^{-1}	ppm (CaCO$_3$)		Ion	mg l^{-1} (ppm)	meq l^{-1}	ppm (CaCO$_3$)
←TH→	Ca^{2+}	80.7	4.04	202	←ALK→	HCO$_3^-$	92.7	1.52	76
	Mg^{2+}	14.0	1.16	58					
	TH	—	5.20	260	←TA→ ←EMA→	SO$_4^{2-}$	135.5	2.82	141
←TC→						Cl$^-$	85.1	2.40	120
	Na$^+$	42.1	1.82	91		NO$_3^-$	30.9	0.50	25
	K$^+$	8.5	0.22	11					
	Total	—	7.24	362		Total	—	7.24	362
←Trace→	Fe	0.07	—	—		CO$_2$	2.9	0.066	3.3
	Mn	0.002	—	—		SiO$_2$	5.3	0.088	4.4
	Al	0.024	—	—		TOC (organic)	2.52 (as C)	—	—

Abbreviations: TH = total hardness, TC = total cations, TA = total anions, ALK = total alkalinity, EMA = equivalent mineral acidity, TOC = total organic carbon

Relationships: $TH + (Na + K) = ALK + EMA$
$TC = TA$

column of highly regenerated strongly acidic cation exchange resin in the hydrogen form. All the cations present will exchange for hydrogen ions which are partly neutralized by basic anions (alkalinity) present to give carbon dioxide. This initially remains dissolved in the column effluent as 'carbonic acid', for example:

$$H^+ + HCO_3^- \rightleftharpoons H_2O + CO_2 \qquad (8.4)$$

$$2H^+ + CO_3^{2-} \rightleftharpoons H_2O + CO_2 \qquad (8.5)$$

Once all basic anions are neutralized the remaining hydrogen ions arising from the cation exchange appear in the effluent with unaffected neutral anions such as chloride, sulfate, and nitrate. Therefore an acidic effluent results which if titrated gives an acid concentration which is equal to the total chemical equivalent concentration of neutral or non-basic anions. The equivalent acid concentration so determined is called the *equivalent mineral acidity*, *EMA*, and is clearly equal to the sum of the equivalent concentrations of sulfate, chloride, and nitrate. Of course, the acidity due to dissolved carbon dioxide is too weak to titrate using methyl orange or an equivalent indicator changing colour at around pH 3.8–4.5.

Reference to Table 8.3 illustrates some fundamental analytical relationships which are all important in assessing the anticipated ion loadings on resins in water treatment applications. The special consideration of expressing carbon dioxide, silica, and organics as 'anions' will be addressed later when discussing demineralizing by ion exchange. Before leaving the subject of water analysis it should be stressed that the principle of measuring cation and anion loadings attributable to electrolytes by displacement of hydrogen ions across a strong cation resin column may be applied to any influent solution provided that no precipitates or insoluble gaseous products are produced.

BOX 8.1 The Determination of Total Anions and Cations in a Solution

By estimating the free acidity or alkalinity of a solution and then estimating the acidity after the solution has passed through a hydrogen form strong cation resin, the total ionic concentration of the solution may be estimated in terms of metal and acid contents.

Requirements

Glass column, 12 mm diameter (approx.)
25 ml of SAC resin (Na or H form)

2 N HCl
Tap water
0.1 N HCl
0.1 N NaOH
Indicator, methyl orange or BDH 4.5

Procedure

1. Titrate 100 ml of tap water with 0.1 N HCl using either indicator. This gives the free alkalinity.
2. Regenerate the resin with 5 BV of 2 N HCl and rinse at 1 BV/5 minutes with deionized water until acid free.*
3. Tap water for analysis is run through the resin at a flow rate of 1 BV/10 minutes. The eluate is collected in 100 ml fractions which are titrated with 0.1 N NaOH using either indicator until two successive samples give an equal titre. The first fraction will always give a smaller titre than the succeeding ones due to dilution caused by the rinse water.

Free Alkalinity

Let:

$$\text{aliquot volume} = A$$
$$\text{normality HCl} = N$$
$$\text{titre HCl} = T$$

$$\text{Free alkalinity} = \frac{1000\,TN}{A} \text{ meq l}^{-1}$$

Total Acidity

Let:

$$\text{aliquot volume} = A$$
$$\text{normality NaOH} = N$$
$$\text{titre NaOH} = T$$

$$\text{Total acidity} = \frac{1000\,TN}{A} \text{ meq l}^{-1}$$

The total acidity value plus the free alkalinity value gives the total metal content of the water.

This method of analysis can be applied to any solution, provided it does not produce insoluble precipitates or gaseous products. Concentrated solutions must be diluted before passing them through the resin. If the solution has a free acidity to the indicator, this is subtracted from the total acidity to give the metal content.

> TIME REQUIRED – 1 hour 30 minutes
>
> * *Note*: If H form resin is available, then only a thorough rinse is required.

SOFTENING

Water softening on originally zeolites and subsequently resins represents the longest established industrial ion exchange unit operation. Water containing greater than 0.4 eq m^{-3} [20 ppm(CaCO$_3$)] total of calcium and magnesium ions is termed 'hard' in that it will not form a stable lather with fatty acid derived soaps. Instead a 'scum-like' product is formed by a precipitation reaction between the organic esters constituting the soap and the calcium and magnesium ions. The action of heat on hard water causes not only the precipitation of calcium carbonate through decomposition of the alkaline hardness but also deposition of non-alkaline hardness salts due to their solubility being exceeded as the hot solution concentrates or the water evaporates from hot surfaces.

Therefore such applications as bottle cleansing, laundering, dyeing, hot water circulation systems, heat exchange circuits, and steam raising by low pressure boilers (2000–4000 kPa) all invariably require softened water. Membrane desalination of hard waters by reverse osmosis often requires the water to be pretreated by softening to avoid scaling of membrane surfaces. Complete softening may be achieved by sodium cycle ion exchange on single fixed beds of strongly functional cation exchange resins operated in either coflow or counterflow mode, the basic process criteria being:

Resin: Any strongly functional styrenesulfonate cation exchange resin. Gel resins (8–12% DVB) are usually used but macroporous types could be preferred under aggressive mechanical conditions, at elevated temperatures, or if oxidizing agents are present.

Ionic Load: Total Hardness (TH) = [Ca + Mg]

Loading: Sodium form column operation at a linear velocity of 5–80 m^3 m^{-2} h^{-1}, whereby hardness cations displace sodium according to the ion exchange reaction:

$$2\,\text{RNa}^+ + \begin{pmatrix} \text{Ca}^{2+} \\ \text{Mg}^{2+} \end{pmatrix} \rightarrow \text{R}_2 \begin{pmatrix} \text{Ca}^{2+} \\ \text{Mg}^{2+} \end{pmatrix} + 2\,\text{Na}^+ \qquad (8.6)$$

The exchange isotherm for calcium–sodium exchange is shown in Figure 8.1 (loading) which demonstrates extremely favourable equilibria towards the removal of calcium from dilute solution. It can be seen that a resin, even when 95% exhausted by calcium (point T), is in equilibrium with product water containing only 5% calcium for an initial total hardness of 0.005 keq m^{-3}. For practical column operations this means that the outlet region of the resin bed can approach high calcium loadings before significant leakage of hardness occurs.

Regenerant: Regeneration is carried out using sodium chloride at a level of around 2 keq(NaCl) m^{-3}$_R$ and concentration 1.7–2.6 keq m^{-3} over a period of 20–30 minutes. The regeneration equilibria are much less unfavourable than might be otherwise anticipated because of the 'electroselectivity' effect (Chapter 5) as depicted by Figure 8.1 (regeneration). In practice regeneration efficiencies of 60–70% are achieved with an average resin conversion of around 55% to the sodium form. A typical economic softening and regeneration equilibrium cycle is represented by curve PQRS in Figure 8.1 which yields operating capacities of around 1.1 keq m^{-3}$_R$.

Softened Water Quality: Treated water hardness values under coflow operation are typically about 1% of the feed total hardness which can

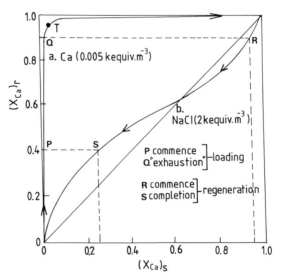

Figure 8.1 *Softening cycle exchange isotherm*
 a) Loading
 b) Regeneration

be improved by employing high regeneration levels (Figure 8.2) or counterflow regeneration. For natural waters of less than 500 mg l^{-1} dissolved solids the $Ca^{2+}-Mg^{2+}-Na^+$ equilibrium is so favourable towards the divalent ion that hardness 'leakage' by sodium elution does not present a serious problem, which is very different from the 'leakage' characteristics encountered during demineralizing processes discussed later. Trace polyvalent cations such as iron(II), iron(III), manganese(II), and aluminium(III) are very strongly taken up by strongly functional cation resins to an extent that their removal during normal regeneration is ineffective. In fact, concentrations of such species in excess of 0.2 mg l^{-1} will slowly irreversibly foul the resin such that consideration may have to be given to their removal by pretreatment.

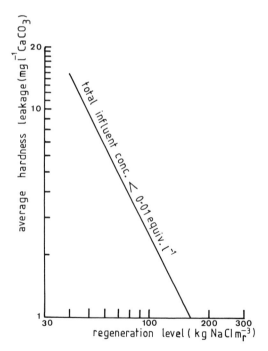

Figure 8.2 *Hardness leakage curve for the softening cycle (coflow)*

BOX 8.2 Softening or Displacement of a Monovalent Ion by a Divalent Ion

While displacement is the more correct term, the best known application is that of softening. A cation resin in the sodium form is used to soften tap water. The residual hardness is tested by EDTA titration or standard soap solution.

Requirements

Glass column, 12 mm diameter (approx.)
25 ml of SAC resin (Na form)
0.1 M ethylenediaminetetraacetic acid disodium salt (EDTA) and eriochrome black T indicator.
 or
Standard soap solution

Procedure

1. Put the resin into the glass column.
2. Pass tap water through the column at 1 BV/5 minutes.
3. Titrate the resulting softened water with EDTA or soap solution and compare with untreated tap water.

Results

An EDTA titration on the tap water will indicate the presence of calcium and magnesium ions in the tap water, the eriochrome black T indicator changing colour from red to blue as these ions are complexed out of solution. In soft water, the eriochrome black T will be blue, since calcium and magnesium ions are absent.

TIME REQUIRED – 30 minutes

Saline Water Softening: For hard alkaline saline waters the high sodium ion concentration results in a far less favourable equilibrium towards calcium and magnesium ions over sodium in which case a need for multistage counterflow operations using softened brine for regeneration is required to achieve an adequate performance. It is largely for these reasons that ion exchange softening of seawater on sulfonate resins, whilst possible in principle, demonstrates inferior economics compared with alternative treatments.

A different situation arises with weakly acidic carboxylate cation exchange resins operating on the sodium cycle. The high selectivity shown by weak carboxylate functional groups towards calcium and magnesium results in effective softening even in the presence of a high sodium ion concentration. A double regeneration technique has to be employed whereupon dilute acid is used to displace calcium and magnesium from the resin followed by treatment with dilute sodium hydroxide to return the resin to the sodium form.

Application in the Chlor–Alkali Industry: The electrolysis of hot saturated brine solution for the manufacture of chlorine gas and sodium hydroxide has advanced in recent years so as to currently favour membrane electrolysis cells rather than the traditional, but environmentally unfriendly, mercury cathode cell. The membrane materials are ion exchange membranes incorporating sulfonic or carboxylic functional groups operating in the sodium form whose function is to selectively transport sodium ions in the brine from the anodic compartment to the cathodic side of the cell where the electrode reactions shown in Figure 8.3 take place.

In order to maintain good ion transport efficiency and acceptable membrane life it is essential that the brine feed contains no greater than $0.05\,\mathrm{mg\,l^{-1}}$ total of calcium and magnesium ions. Conventional ion exchange softening of even chemically presoftened brine containing $2-10\,\mathrm{mg\,l^{-1}}$ hardness cannot economically produce the final quality required against a background concentration of some $310\,000\,\mathrm{mg\,l^{-1}}$ sodium chloride. Instead advantage is taken of chelating resins based on iminodiacetate, $-\mathrm{N[CH_2CO_2^-]_2}$, or aminophosphonate, $-\mathrm{NHCH_2PO_3^{2-}}$, functionality which exhibit very favourable selectivity towards calcium and magnesium cations (see Chapter 2, 'Special Ion Exchange Materials').

Lead and polishing columns are operated in series to treat hot alkaline brine (50–80 °C) with the resin in the sodium form. A double regeneration technique is employed whereby dilute hydrochloric acid is used to displace the calcium and magnesium from the resin followed by conversion of the resin to the sodium form using dilute sodium hydroxide. The hot conditions together with the appreciable resin volume changes that occur when cycling between alkaline–acid conditions demand that resins of macroporous structure are used.

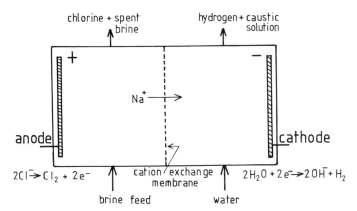

Figure 8.3 *The Chlor–Alkali Membrane Electrolysis cell*

DEALKALIZATION

Dealkalization by ion exchange usually employs weakly acidic carboxylic acid resins operating on the hydrogen cycle. The high selectivity shown towards divalent calcium and magnesium over monovalent sodium results in the preferential exchange of calcium and magnesium ions (denoted M^{2+}) thus:

$$2\,RCO_2H + M^{2+} \rightarrow (RCO_2)_2M + 2\,H^+ \qquad (8.7)$$

In the presence of hydrogen ions the above exchange would *not occur* (Chapter 4, 'Resin Description') but if alkaline hardness is present the exchanged hydrogen ions are immediately neutralized by the basic hydrogencarbonate and carbonate anions present to give carbon dioxide which remains dissolved in the water as weak 'carbonic acid' as given by equations 8.4 and 8.5.

The exchange of calcium and magnesium ions continues until all basic anions are neutralized after which time no further exchange can occur and therefore the extent of hardness removal is equivalent to the alkaline hardness of the water. Typical operating parameters for dealkalization are:

Resin: Any weak carboxylic acid cation exchange resin (WAC).

Ionic Load: Total Alkalinity = *M.ALK*

Loading: Hydrogen form column operation at typically linear velocities of 5–80 $m^3\,m^{-2}\,h^{-1}$, but since slower particle diffusion kinetics prevail the operating capacity is both flow and temperature dependent. In other words the resin is rate sensitive such that for a given influent alkalinity the operating capacity decreases with increasing loading rate and decreasing temperature as illustrated by Figure 8.4. The operating capacity for total dealkalization on weakly functional cation exchange resins is reduced if sodium alkalinity is present, *i.e.* *M.ALK* > *TH*, owing to the lower selectivity for sodium compared with divalent cations. In such a case, which is reasonably rare for natural water supplies, the operating capacity for alkaline hardness uptake remains unaffected but sodium alkalinity is displaced relatively quickly.

Regenerant: Regeneration is carried out with either dilute hydrochloric acid or dilute sulfuric acid at concentrations of 1.4 $keq\,m^{-3}$ and 0.16 $keq\,m^{-3}$ respectively over a period of about 30 minutes.

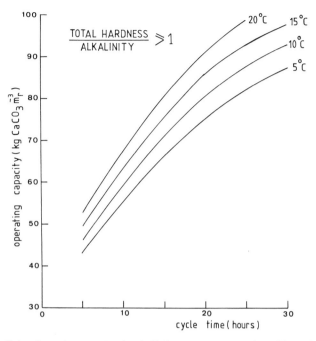

Figure 8.4 *Operating capacity for dealkalization on a typical weakly acidic cation exchange resin at various loading rates and water temperatures*

Being a weak acid exchanger the resin is regenerated at virtually 100% efficiency, giving an operating capacity nearly equivalent to the regeneration level employed up to a maximum given by the exhaustion capacity for the resin for a given flowrate. If sulfuric acid is used as regenerant it is essential that its concentration does not exceed 0.16 keq m^{-3} otherwise there is a very real risk of precipitating calcium sulfate within the resin bed or even within the resin beads themselves. Even at the sulfuric acid concentration cited the equivalent concentration of calcium sulfate released exceeds its theoretical solubility but fortunately remains in supersaturated solution for the duration of the regeneration cycle provided injection and rinse flows are not interrupted.

Figure 8.5a shows the acid injection and breakthrough profiles for the regeneration of an industrial dealkalization unit, where it may be observed that no immediate acid breakthrough occurs giving essentially a neutral regeneration effluent for most of the injection cycle. Compare this with Figure 8.5b showing the acid regeneration profile for a strong acid cation resin where breakthrough occurs immediately giving coflow efficiencies of typically about 40%.

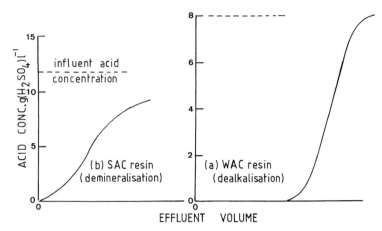

Figure 8.5 *Regeneration breakthrough profiles for weak and strong acid cation exchange resins*

Treated Water Quality: The residual alkalinity in the product water typically follows the profile shown in Figure 8.6 ranging from near zero (pH 3.8–4.0) at the beginning of the run up to about 10% of the influent alkalinity at exhaustion of the resin (pH 5.6–5.8). If the quantity of regenerant used is less than that equivalent to the exhaustion capacity of the resin then it is essential to meter the service cycle to a strict volume throughput equivalent to a capacity calculated from the quantity of regenerant employed.

Sometimes slight acidity in the treated water may be observed very early in the service cycle which is due to most weak acid resins

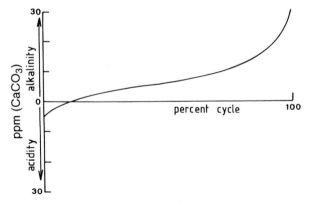

Figure 8.6 *A typical alkalinity leakage profile during dealkalization on a weak acid carboxylate cation exchanger*

possessing a small salt splitting capacity (Chapter 4, 'Chemical Specification'). A prolonged duration of acid water is an indication that the resin has been over-regenerated thereby fully regenerating the fraction of stronger carboxylic acid groups within the resin that would otherwise remain exhausted if the correct stoichiometric amount of acid regenerant is used. It is usual for the dissolved carbon dioxide produced during the dealkalizing process to be degassed (DG) by spraying the water down a packed tower against an upflow of air which strips out the carbon dioxide to give a residual typically less than $5 \text{ mg}(CO_2) \text{ l}^{-1}$.

Dealkalization and alkaline hardness removal are synonymous terms which means that after such a treatment the water is partially softened. Full softening may be accomplished through further treatment by conventional sodium cycle softening to remove the remaining permanent hardness. However, one very significant difference applies to water softened directly by the sodium cycle on strongly functional sulfonate resins, compared with that obtained by softening a prior dealkalized–degassed feed. In the former case the cationic content of the treated water is entirely sodium ions together with a total equivalent content of all the anions originally present, and therefore the dissolved solids content of the water remains virtually unchanged. For the latter instance, whereas the cation content of the water is still all sodium, the alkaline hardness has been firstly substituted by an equivalent quantity of carbon dioxide which is degassed and after further softening only neutral salts of sodium remain. Thus the total dissolved solids content of the water has been reduced by an amount equivalent to the alkaline hardness content of the raw water. Therefore a (dealkalized–degassed–softened) water is partially *demineralized* and is often preferred as feed for moderate pressure boiler plant.

A much simplified representation of a boiler is shown in Figure 8.7. As water loss through evaporation occurs (*steaming rate*, $E \text{ kg h}^{-1}$) the boiler water becomes more concentrated in dissolved solids increasing up to a maximum allowable concentration S_B dictated by the boiler design and steam chemistry requirements. In order that this limit is not exceeded it is necessary to discard a portion of the boiler water, termed *blowdown*, $B \text{ kg h}^{-1}$, and to compensate this loss by admitting fresh feedwater or *make-up*, $M \text{ kg h}^{-1}$. The make-up water is supplied either entirely by the ion exchange water treatment plant (*make-up plant*) or supplements the feedwater recovered from condensed clean steam, *condensate return*, $CR \text{ kg h}^{-1}$. In the simplest case of assuming the solids carried over with the steam S_S to be zero and taking a mass

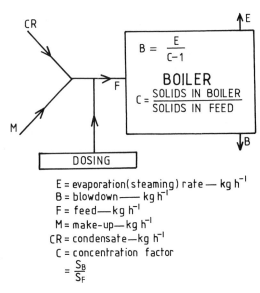

Figure 8.7 *A simple representation of mass balance and blowdown for a boiler*

balance on water and solids, the boiler blowdown required to maintain the boiler dissolved solids at S_B is given by $B = E/(C - 1)$, where C is the concentration factor given by (S_B/S_F), allowance being made for dilution of the make-up water dissolved solids with condensate.

The dissolved solids quality of the boiler feed S_F is governed by the design and performance of the water treatment plant and the level of *chemical dosing* which is required to maintain the correct system chemistry. If either source is a cause of excess dissolved solids the concentration factor reduces resulting in increased blowdown for a given control on total dissolved solids in the boiler *TDS* and steaming rate. Besides possible boiler damage through solids deposition and corrosion, the cost of excessive blowdown arising from inferior boiler feed quality is severe since the replacement of hot blowdown with fresh boiler feed represents an energy loss from the system. This has to be made up at the cost of extra fuel consumption in raising further 'cold' make-up to the operating temperature of the boiler.

The partial demineralization of water by the dealkalization–degassing–softening route may be described as a combined cycle process meaning that two different regenerants are used. Dealkalization by itself is a single cycle process and is widely used to remove alkaline hardness from waters used for cooling, and in the beverage industries where uncontrolled alkaline hardness can deleteriously affect the quality of product.

Before leaving the topic of dealkalization on weak acid resins it is interesting to note that prior to the advent of carboxylic acid resins alkalinity removal was achieved on sulfonated coal exchangers (Chapter 2, 'Early Organic Ion Exchange Materials'). Use was made of their low but useful weak carboxylic acid functionality and stoichiometric regeneration with mineral acid. Although obsolete as commercial ion exchangers, ongoing investigations by Streat and Nair show that oxidized forms of related carbonaceous materials (activated carbons) have a potential use as sorbents of heavy metal pollutants by ion exchange fixation on carboxylic acid groups.

OTHER SINGLE CYCLE ION EXCHANGE PROCESSES IN WATER TREATMENT

Organic Removal

Surface water abstracted from catchments accessible to run-off collected from rural land masses nearly always contains significant levels of natural organic matter. It remains impossible to categorize the specific chemical structure of the organic species but they are essentially colloidal derivatives of humic and fulvic acids arising from decaying vegetation and peaty soils. Figure 8.8 illustrates one suggested structure for such compounds from which it may be appreciated that one is dealing with macromolecules of molecular mass in the region of 500 to > 100 000 and of size 10–1000 nm (see Figure 8.9).

The concentration of the various 'organics' present in surface waters are not amenable to separate absolute measurement and instead they are estimated collectively by standard methods. These include oxidation with acidified potassium permanganate and reported as oxygen absorbed in $mg(O_2) \, l^{-1}$, ultraviolet absorbance at a wavelength of 300 nm, or the now preferred oxidative destruction of the organics to carbon dioxide which is the basis of reporting 'organics' as $mg(carbon) \, l^{-1}$, i.e. $mg(C) \, l^{-1}$, otherwise called total organic carbon (TOC). Typical TOC values for UK raw surface waters can be expected to lie in a range of $2-20 \, mg(C) \, l^{-1}$ and is seasonal, commonly peaking in the Autumn and Spring, and loosely coincident with higher rainfall during these periods as shown in Figure 8.10.

The exact nature of the organic matter is ill-defined but it is known to behave as a colloid and is often complexed with silica and heavy metal atoms such as iron, aluminium, and manganese. The weak carboxylic acid functionality of 'humus' organics means that much of

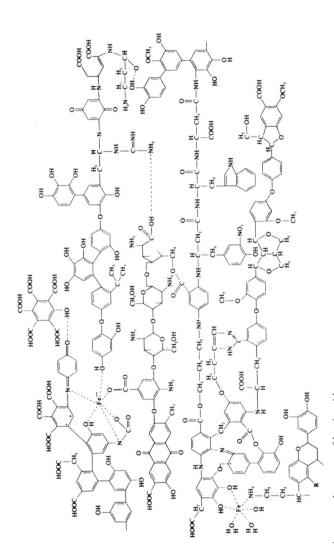

Figure 8.8 *A suggested structure of humic acid* (Reproduced by permission from G. M. Tilsley, 'Interaction of Organic Matter with Anion Resins', *Chem. Ind. (London)*, 1979, No. 5, 142.

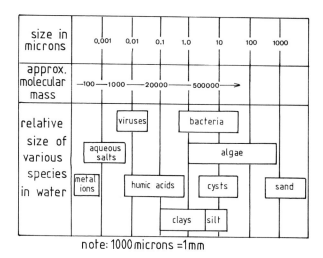

Figure 8.9 *A spectrum of particulate sizes encountered in the aqueous environment*

the organic matter behaves as a polycharged macro-anion capable of being taken up by anion exchange resins, which is the reason why 'organics' are listed as anions in Table 8.3. The large complex structure of such ions results in their possibly becoming irreversibly entangled or bound to the resin matrix through a variety of complex interactions (Chapter 5), thereby blocking exchange sites causing the resin to lose operating capacity and become rate sensitive.

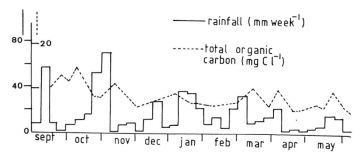

Figure 8.10 *Frequency plots of the total organic carbon (TOC) levels and incident rainfall for a scottish upland water*
(Data from J. W. Parsons, M. G. Kibblewhite, B. B. Sithole, and E. H. Voice, International Conference on Water Chemistry of Nuclear Reactor Systems, BNES, Bournemouth, 1980)

In order to minimize the risk of encountering organic fouling problems in ion exchange processes using anion exchange resins it is often appropriate to employ a strongly basic anion exchange resin operating on a coflow chloride cycle as an *organic trap* for the removal of organics. Such a process is not as sacrificial on the anion exchange resin as might be thought since an anion resin possessing one or both of the properties of macroporosity and an acrylic matrix shows good reversibility towards the exchange of organics, such that a resin life of some 3–4 years is not uncommon. In fact experience has shown that using a combination of resins showing different structural properties, gel and macroporous, styrenic and acrylic, either in the same or separate columns can effect better organic removal than a single resin alone. It would seem as if a 'cocktail' of resin structures is better able to treat the broad molecular spectrum which constitutes a typical distribution of natural organics.

The operating characteristics of an *organic trap* or *organic scavenger* unit are generally as follows:

Resin: A single or mix of Type 1 strongly basic anion exchange resins selected on the basis of their proven resistance to irreversible fouling.

Ionic Load: Total Organic Carbon (TOC) or some other equivalent assessment.

Loading: 5–80 $m^3 m^{-2} h^{-1}$ to a strict metered capacity governed by load and flowrate.

Regenerant: Sodium chloride solution of concentration 1.7 $keq\, m^{-3}$ applied at a level of approximately 3 $keq(NaCl)\, m^{-3}{}_R$ over a period of 30–60 minutes. It is preferable to make the regenerant alkaline with sodium hydroxide (0.5 $keq\, m^{-3}$) which solubilizes humic organic matter making it easier to elute off the resin.

Treated Water Quality: Typically about 50–70% removal of organics depending upon the nature of the species present. It is not essential for organic removal that the resin should possess a high ion exchange capacity and in fact it is more desirable that they possess greater porosity even at the expense of capacity. However, whatever the available exchange capacity, the raw water will be changed in composition due to ion exchange across the chloride form of the resin.

Consider a typical effluent (treated water) composition profile during loading for chloride cycle exchange as shown in Figure 8.11.

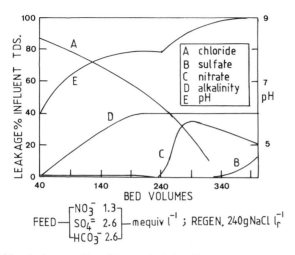

Figure 8.11 *Leakage profiles of ion residuals for chloride cycle anion exchange on a strongly basic (Type 1) styrenic resin*
(Reproduced by permission from F. X. McGarvey and R. Gonzalez, 'Ion Exchange Studies on Strongly Basic Anion Exchange Resins Prepared with Tertiary Amines of Varying Molecular Weight', in 'Ion Exchange Advances', ed. M. J. Slater, Elsevier Applied Science, London and New York, 1992, p. 97)

Whilst sorption of organics takes place all the anions in the water are exchanging for chloride ions; therefore *initially* the effluent is virtually all chloride of concentration equal to the total anion concentration. As the run continues hydrogencarbonate ion is displaced early on by exchange for chloride whilst sulfate and nitrate are favourably retained. Eventually nitrate ion breakthrough is observed at an increasing concentration governed largely by the ratio of sulfate to nitrate concentration in the influent. Thus the overall breakthrough profile is in keeping with the normal ion selectivity sequence for strong base Type 1 anion exchange resins (Chapter 5, 'Dilute Solution Anion Exchange'). Organic removal by ion exchange is very often a pretreatment stage before other downstream ion exchange operations and therefore it is important to take into account the effect of the changing effluent composition which becomes the feed for downstream processes.

In theory, chloride cycle anion exchange could be used for dealkalization but the efficiency is low (approx. 25%), gives a low operating capacity (approx. 0.3 keq m^{-3}), and does not give a reduction in total dissolved solids.

Nitrate Removal

The quality of potable water supplies and water used for production of foods and beverages is required to meet the EC directives on water quality which for nitrates (Table 8.2) specifies the maximum allowable concentration to be 50 mg(NO_3^-) l^{-1}. Over many years the water abstracted from certain sources, particularly groundwaters, has risen to above the allowable limit. Whilst the problem may be overcome by blending low and high nitrate sources, conservation of supplies and limitations in water transfer options has meant that it is now necessary to treat some waters at source for nitrate reduction. Biological denitrification and ion exchange are both explored process options with ion exchange being the easiest and most economic expedient at the present time.

The process is very similar to that for organic removal, being a single cycle ion exchange treatment on the chloride cycle. With conventional strongly basic anion exchange resins the residual ion concentrations follow a profile not unlike that shown in Figure 8.11 which highlights the following consequences. Firstly it may be observed that initially nitrate is reduced very effectively as are all other ions giving essentially a water very high in chloride. Later in the cycle the chloride residual reduces as hydrogencarbonate breaks through whilst the sulfate and nitrate are still retained. Finally, the nitrate residual rises sharply due to displacement by sulfate.

Thus over the whole cycle the quality of the treated water is changing drastically with respect to chloride and alkalinity which presents corrosivity control difficulties with regard to the downstream distribution network. Also if the sulfate/nitrate concentration ratio in the influent is high ($\gg 1$), an overrun of the column may cause effluent nitrate residuals to quickly rise to *above* their influent concentration due to displacement by a high concentration front of sulfate ion.

To overcome these problems use is made of the now available *nitrate-selective* resins (Chapter 5, 'Dilute Solution Anion Exchange') for which the anion affinity sequence is:

$$NO_3^- > SO_4^{2-} > Cl^- > HCO_3^-$$

Now the exchange profile takes the form of Figure 8.12, where hydrogencarbonate and sulfate break through relatively early and the selectivity inversion for NO_3^-/SO_4^{2-} means that residual nitrate levels cannot rise above the influent concentration. The overall result is a

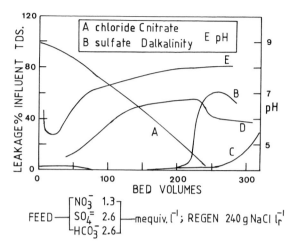

Figure 8.12 *Leakage profile of ion residuals for chloride cycle anion exchange on a 'nitrate selective' resin*
(Reproduced by permission from F. X. McGarvey and R. Gonzalez, in 'Ion Exchange Advances', ed. M. J. Slater, Elsevier Applied Science, London and New York, 1992, p. 97)

water low in nitrate and of a much more acceptable mineralogical composition. Coflow and counterflow designs are employed depending upon the residual nitrate levels required. Regeneration is carried out using sodium chloride solution at levels of around 4 keq(NaCl) m$^{-3}_R$ and 2 keq(NaCl) m$^{-3}_R$ respectively giving operating capacities in the range 0.3–0.6 keq m$^{-3}_R$ with respect to nitrate. Some designs allow for a further regeneration of the resin with dilute sodium hydrogencarbonate solution which has the effect of raising the residual hydrogencarbonate concentration (alkalinity) in the treated water thus minimizing its corrosivity towards the distribution pipework. It is important to realize that the regeneration effluents are very concentrated in nitrates, requiring carefully controlled disposal so as they do not re-enter fresh water courses. One very interesting development idea is that of combined ion exchange and biological–chemical denitrification, the latter process being used to denitrify the effluent for reuse as regenerant.

Oxygen Removal

A long-established process for removing dissolved oxygen from otherwise demineralized water is by passage through a strong base anion exchange resin operating on the sulphite cycle as represented by the equations:

$$\text{Service} \quad 2\,R_2SO_3 + (O_2)_{aq} \rightarrow 2\,R_2SO_4 \quad (8.8)$$

$$\text{Regeneration} \quad R_2SO_4 + Na_2SO_3 \rightarrow R_2SO_3 + Na_2SO_4 \quad (8.9)$$

DEMINERALIZATION

Demineralization is a combined cation (hydrogen) and anion (hydroxide) exchange cycle process to produce 'pure' water. The need for demineralized water in the process and power generation industries is vast. The quality of water required is varied depending upon the application but, in terms of electrical conductivity, which is a measure of the ionic impurity concentrations in a water, industrial requirements range from typically < 10 microSiemens cm^{-1} (μS cm^{-1}) for many metal and textile finishing processes to ultimate ionic purity (conductivity = 0.04 μS cm^{-1} at 18 °C) required by the electronics industry. Quality may be impaired by non-ionic impurities present which do not contribute to conductivity, and this is discussed briefly in later text. It is somewhat overlooked, not to say taken for granted, that modern ion exchange practices would regard a demineralized water quality requirement of conductivity 0.1 μS cm^{-1} as commonplace which in ion impurity terms translates to around 99.999998% pure water.

Consider a series two-stage process where raw water is passed firstly through a strong acid cation exchange resin in the hydrogen form, the effluent from which is then passed through a column of strongly basic anion exchange resin in the hydroxide form. Across the cation exchange column all cations exchange for hydrogen ions to give a dilute acidic effluent made up of acid sulfates, nitrates, and chlorides together with dissolved carbon dioxide. Upon passing through the anion exchange column, neutralization of the acid feed occurs through the exchange of all anions for hydroxide to give 'pure water'. The ion exchange reactions may be represented in an idealized way thus:

$$RH + \text{(all cations)} \rightarrow R(\text{cations}) + H^+ \quad (8.10)$$

$$ROH + \text{(all anions)} \rightarrow R(\text{anions}) + OH^- \quad (8.11)$$

$$H^+ + OH^- \rightleftharpoons H_2O \quad (8.12)$$

In practice the exchange reactions depicted above are not complete owing to the 'leakage' of residual ions as described previously in Chapter 7 (Column Breakthrough and 'Leakage').

COFLOW TWO-STAGE SYSTEMS

Strong Acid Cation–Strong Base Anion

Consider coflow two-stage demineralization using strongly functional cation and anion exchange resins in the hydrogen and hydroxide forms respectively as illustrated in Figure 8.13. The operating requirements of this process are as follows:

Cation Column

Cation Resin: Any styrenesulfonic strong acid resin in the hydrogen form (SAC). Most applications call for gel resins unless high differential pressures, resin transfers, or aggressive chemical conditions suggest a preference for macroporous products. Standard bead size gradings are usually adequate, but a more uniformly graded coarse bead size may be preferable for deep beds (> 1.5 m). Much interest is currently being shown in the recently available 'monosized' resins because of their predictable hydrodynamic behaviour and possibly as

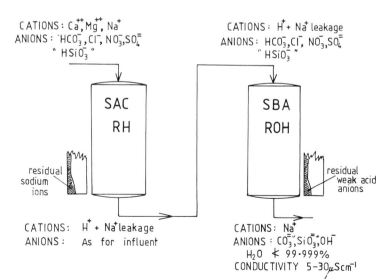

Figure 8.13 *Two-stage demineralization (coflow regeneration)*

a means of improving regeneration efficiency and rinse characteristics through their faster kinetics.

Ionic load: Total Cations $(TC) = ALK(M) + EMA$

Loading: Hydrogen form column operation at typically 5–80 m³ m^{-2} h^{-1} linear flowrate.

Regenerant: Dilute hydrochloric acid (1.4 keq m^{-3}) or sulfuric acid (0.3–1.0 keq m^{-3}), the latter concentration being decided by the calcium content of the raw water thereby avoiding calcium sulfate precipitation. Regeneration levels are typically 1.3–2.0 keq m^{-3}$_R$ injected over a period of about 30 minutes.

Cation Column Effluent Quality: As the water passes down the column stoichiometric ion exchange occurs to give a dilute mixture of acids as given by the following reactions:

$$RH + (\tfrac{1}{2}Ca^{2+}, \tfrac{1}{2}Mg^{2+}, Na^+) \begin{bmatrix} HCO_3^- \\ \tfrac{1}{2}SO_4^{2-} \\ Cl^- \\ NO_3^- \\ HSiO_3^- \\ organics \end{bmatrix} \rightarrow R\begin{bmatrix} \tfrac{1}{2}Ca \\ \tfrac{1}{2}Mg \\ Na \end{bmatrix} + H^+ \begin{bmatrix} HCO_3^- \\ \tfrac{1}{2}SO_4^{2-} \\ Cl^- \\ NO_3^- \\ HSiO_3^- \\ organics \end{bmatrix}$$

(8.13)

Also shown in Figure 8.13 are the post-coflow regeneration exchange site distributions existing at the outlet regions of the cation and anion resin columns. Therefore in the case of the cation unit as the acid effluent emerges from the column, residual sodium ions are eluted off the resin, and appear in the product as sodium leakage (Chapter 7, 'Column Breakthrough and "Leakage" '):

$$\underset{\text{residual sodium sites}}{RNa} + H^+ \rightarrow RH + \underset{\text{"leakage"}}{Na^+} \quad (8.14)$$

The titratable acidity at the cation outlet is called the *Free Mineral Acidity (FMA)* and is related to the Equivalent Mineral Acidity (*EMA*) by the relationship:

$$FMA + \text{Leakage} = EMA$$

Other monovalent cations of low selectivity for the resin such as potassium ion (K^+) and ammonium ion (NH_4^+) would show similar leakage characteristics. For UK waters a typical average sodium leakage would be about $1\ mg(Na)\ l^{-1}$, increasing with increasing sodium fraction of the total cations.

Anion Column

Anion Resin: Any strongly basic Type 1 or Type 2 quaternary ammonium resin (SBA) in the hydroxide form. The slightly weaker basicity of the Type 2 products results in achieving a higher operating capacity (approx. $0.7\ keq\ m^{-3}{}_R$). As a consequence of the higher operating capacity, and lower basicity, the Type 2 resins usually perform at a higher silica leakage than the Type 1 exchangers given the same operating conditions.

If treated water quality with regard to silica is non-critical, or is removed at a later stage, there are significant operating cost benefits in employing Type 2 resins but this has to be set against their more rapid capacity loss, and therefore shorter operational life compared with Type 1 materials (Chapter 4).

Ionic Load: Total Anions $(TA) = ALK(M) + EMA +$ free $CO_2 +$ Silica

Loading: Hydroxide form column operation at typically $5-80\ m^3\ m^{-2}\ h^{-1}$ linear flowrate.

Regenerant: Dilute sodium hydroxide ($1.0-1.2\ keq\ m^{-3}$) at a level of typically $1.6-2.0\ keq(NaOH)\ m^{-3}{}_R$ over a period of about 30 minutes.

Treated Water Quality: The influent total anion concentration is present as dissolved carbon dioxide (equivalent to the original total alkalinity plus original free dissolved carbon dioxide), plus mineral acids (equivalent to the *EMA*), plus reactive 'silica'. The exact manner by which 'silica' (silicon(IV) dioxide) is taken up by a strongly basic anion exchange resin is known to be complex; but it is best regarded as the hydrogensilicate ion ($HSiO_3^-$), and like dissolved carbon dioxide, is exchanged by neutralization of the appropriate weak acids:

$$SiO_2 + H_2O \rightleftharpoons H_2SiO_3 \rightleftharpoons H^+ + HSiO_3^- \qquad (8.15)$$

$$ROH + HSiO_3^- + H^+ \rightarrow RHSiO_3 + H_2O \qquad (8.16)$$

$$CO_2 + H_2O \rightleftharpoons H_2CO_3 \rightleftharpoons H^+ + HCO_3^- \qquad (8.17)$$

$$ROH + HCO_3^- + H^+ \rightarrow RHCO_3 + H_2O \qquad (8.18)$$

Therefore the overall ion exchange neutralization reactions occurring across the anion resin column may be represented collectively by the following general scheme:

$$ROH + (H^+) \begin{bmatrix} HCO_3^- \\ HSiO_3^- \\ \tfrac{1}{2}SO_4^{2-} \\ Cl^- \\ NO_3^- \\ \text{organics} \end{bmatrix} \rightarrow R \begin{bmatrix} HCO_3 \\ HSiO_3 \\ \tfrac{1}{2}SO_4 \\ Cl \\ NO_3 \\ \text{organics} \end{bmatrix} + H_2O \qquad (8.19)$$

In fact the exchange of dibasic acids such as 'carbonic acid' and sulfuric acid is complex. Initially, they are taken up by the basic hydroxide form of the resin as carbonate and sulfate but become converted to the hydrogencarbonate and hydrogensulfate form as the resin exhausts from the top. The cation content of the anion column influent is not solely hydrogen ions since some sodium ions are present as leakage from the cation column which pass unchanged through the anion column. Thus sodium leakage appears in the anion column effluent as a stoichiometric concentration of very dilute sodium hydroxide or *'caustic slip'*. Furthermore the slight concentration of sodium hydroxide elutes post-regeneration residuals of the lowest affinity anions, 'silica' and hydrogencarbonate, from the bottom of the column as represented thus:

$$RHCO_3 + 2\,OH^- \rightarrow ROH + CO_3^{2-} + H_2O \qquad (8.20)$$

$$RHSiO_3 + 2\,OH^- \rightarrow ROH + SiO_3^{2-} + H_2O \qquad (8.21)$$

The combined leakage from both the cation and anion columns therefore results in traces of sodium hydroxide, sodium carbonate, and 'silica' in otherwise demineralized water. Demineralized water from a coflow two-stage Strong Acid Cation–Strong Base Anion plant shows a conductivity profile similar to that given in Figure 8.14 with a typical average conductivity of about 10 microSiemens cm^{-1}, silica at around 0.1–0.3 mg(SiO$_2$) l^{-1} and pH between 9 and 10. At true exhaustion of the strong base anion unit the effluent pH drops with

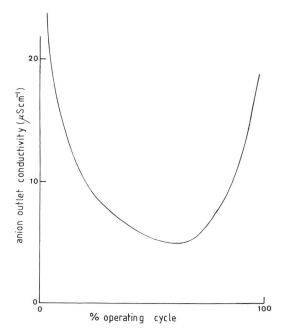

Figure 8.14 *A typical coflow demineralized water conductivity profile*

hydrogencarbonate and 'silica' being displaced as their respective weak acids.

The behaviour of humic organics present in water towards the anion resin is difficult to systematize except that as essentially large high molecular mass polycarboxylate ions they are exchanged onto the anion resin (see earlier section on Organic Removal). Unless the resin shows reversibility to the exchange of such complex ions through a selection of either macroporous structure and or an acrylic matrix (Chapter 3), particularly high burdens of organics may irreversibly foul the resin leading to a deterioration in water quality and the phenomenon of excessively long rinses before acceptable quality is reached. Long rinses are thought to be caused by the presence of organic anions trapped or entangled in the resin matrix, which after caustic regeneration exist in the sodium carboxylate salt form. Subsequent hydrolysis during the rinse causes a prolonged trace of sodium hydroxide to leach into the demineralized water, thus:

$$\text{Resin} \sim \underset{\text{organic salt form}}{(CO_2)_n{}^{n-}(Na^+)_n} + H_2O \rightarrow \text{Resin} \sim \underset{\text{organic acid form}}{(CO_2H)_n} + n\,NaOH$$

(8.22)

Strong Acid Cation–Weak Base Anion

The cation exchange step is exactly as described for the Strong Acid Cation (SAC)–Strong Base Anion (SBA) process but now the acidic cation column effluent passes down a column of weakly basic anion exchange resin. The strong mineral acids are taken up by the anion resin through addition to form the acid salt forms, whilst the too weakly acidic dissolved carbon dioxide and 'silica' pass through unaffected (Chapter 4).

Resin: Any weakly basic anion exchange resin (WBA).

Ionic Load: Equivalent Mineral Acidity = *EMA*

Loading: Free base form operation at a typical flowrate of $5-80 \text{ m}^3 \text{ m}^{-2} \text{ h}^{-1}$. Unlike water treatment service cycles on strongly functional resins which are all film diffusion controlled; the loading cycle on weakly basic anion exchange resins is often particle diffusion controlled and therefore rate sensitive.

Regenerant: Normally dilute sodium hydroxide, but being a weakly basic exchanger less basic regenerants such as sodium carbonate, calcium hydroxide, and ammonium hydroxide may sometimes by used quite effectively. Regeneration with sodium hydroxide is extremely favourable giving efficiencies of the order of 60–80%. However, because the free base form of a weak base resin carries no ionic charge the Donnan sorption of sodium hydroxide can readily occur (See Chapter 5, 'Sorption of Non-exchange Electrolyte and the Donnan Equilibrium'). It is for this reason that weak base resins demonstrate high rinse volumes.

Treated Water Quality: The acid addition reactions undergone by the weakly basic anion exchange resin are shown by the following overall scheme:

$$RN(CH_3)_2 + H^+ \begin{bmatrix} HCO_3^- \\ HSiO_3^- \\ \tfrac{1}{2}SO_4^{2-} \\ Cl^- \\ NO_3^- \\ organics \end{bmatrix} \rightarrow RNH(CH_3)_2^+ \begin{bmatrix} Cl^- \\ \tfrac{1}{2}SO_4^{2-} \\ NO_3^- \\ organics \end{bmatrix} + H_2CO_3 + H_2SiO_3$$

(8.23)

For truly weak base anion resins containing only secondary or tertiary amine groups the ion of least affinity for the resin is chloride which appears as leakage. In fact most modern weak base exchangers contain a fraction of strong base groups which results in the final demineralized water containing traces of sodium alkalinity, as well as sodium chloride and virtually influent concentration of dissolved carbon dioxide and silica. The conductivity of the final water is usually between 5–30 μS cm^{-1} of pH varying from approximately 9–5 across the exhaustion cycle.

BOX 8.3 Two-stage Demineralization of a Solution

In this experiment the demineralization of a solution may be followed by using coloured ions, and observing the formation of coloured bands on the resins and the decolorization of the solution. It is thus particularly suitable for demonstration. A blue cation and yellow anion are included in the solution, and the concentrations are chosen so that a green liquor is obtained for demineralization. Extraction of the blue cation on the SAC resin column is easily seen, and the resulting cation free liquor is yellow. The yellow dichromate is then removed by the SBA resin column and the result is a colourless deionized water.

Requirements

2 glass columns, each 12 mm diameter (approx.)
25 ml of SAC resin (Na form)
25 ml of SBA resin (Cl form)
Solution containing 0.6 g l^{-1} CuSO$_4$.5H$_2$O and 0.08 g l^{-1} K$_2$Cr$_2$O$_7$
1 N HCl
1 N NaOH

Procedure

1. Put the SAC resin into a column and regenerate it with 4 BV of 1 N HCl. Rinse with 7 BV of deionized water at 1 BV/5 minutes.*
2. Put the SBA resin into the second column and regenerate it with 1.5 BV of 1 N NaOH. Rinse with 7 BV of deionized water at 1 BV/5 minutes.*
3. Connect up the bottom of the column containing SAC resin to the top of the SBA resin column, using rubber tubing, so that the effluent from the SAC column is fed into the SBA resin column.
4. Pass the CuSO$_4$/K$_2$Cr$_2$O$_7$ solution, which is greenish yellow in colour, through the resins at a flow rate of 50 ml/15 minutes. The SAC resin removes the potassium and copper, and a green band will form on it due to the copper. The eluate from the SAC column, containing sulfuric and chromic acids, is decolorized by the SBA column. An orange band is formed on the SBA resin due to the chromate.

Conductivity measurements of the effluent after treatment will verify that the solution has been deionized.

$$2\,RH + CuSO_4 \rightleftharpoons R_2Cu + H_2SO_4$$
$$2\,RH + K_2Cr_2O_7 \rightleftharpoons 2\,RK + H_2Cr_2O_7$$
$$2\,R'OH + H_2SO_4 \rightleftharpoons R_2'SO_4 + 2\,H_2O$$
$$2\,R'OH + H_2Cr_2O_7 \rightleftharpoons R_2'Cr_2O_7 + 2\,H_2O$$

TIME REQUIRED – 3 hours

* *Note*: Alternatively, rinsed pre-regenerated H and OH ionic forms may be used.

COFLOW MULTISTAGE PROCESSES

Consider the following multistage processes where:

WAC ≡ Weak acid cation stage
SAC ≡ Strong acid cation stage
WBA ≡ Weak base anion stage
SBA ≡ Strong base anion stage
DG ≡ Degassing stage
⟶ ≡ direction of service flow
---→ ≡ direction of regeneration flow

1. $\;$ SAC − DG − SBA

Here a degassing stage precedes the strong base anion column which lengthens its loading cycle by virtue of the fact that the anionic load is reduced by an amount equivalent to the raw water alkalinity, thereby greatly improving the anion unit operating costs.

2. $\;$ SAC − WBA − DG

It is usual to degas demineralized water produced by weak base anion schemes, but the presence of dissolved carbon dioxide helps suppress the pH across the anion loading cycle thereby assisting ionization of the free base form functional groups. Therefore it is preferred to position the degasser downstream, not upstream, of the weak base anion column especially if the raw water has a low *EMA* value.

3. $\overset{\downarrow}{|\text{WAC}|}\overset{\downarrow}{|}-\overset{\downarrow}{|\text{SAC}|}-\text{DG}-\overset{\downarrow}{|\text{SBA}|}$

Using weak–strong resin combinations of the same charge type, *i.e.* cation or anion, is often a powerful means of greatly improving both the volume throughput between regenerations and column regeneration efficiency, thereby reducing chemical operating costs. Dealkalization across the leading weak acid cation unit reduces the cation load onto the strong acid cation column. Furthermore, the spent acid regenerant from the strong acidic cation column may be used to regenerate the weak acid cation resin – a technique known as *thoroughfared regeneration*. In this way the overall regeneration efficiency across the cation exchange section may often be greatly increased, for waters containing a significant fraction of alkaline hardness.

4. $\overset{\downarrow}{|\text{SAC}|}-\overset{\downarrow}{|\text{WBA}|}-\overset{\downarrow}{|}-\text{DG}-\overset{\downarrow}{|\text{SBA}|}$

The above scheme is the anion exchange analogy of Scheme 3 employing throughfared regeneration of the weak base anion column. Here, the strong base anion resin is acting as a 'polisher' to remove ion species passing the weak base anion unit. An added advantage of this configuration is that a macroporous or acrylic weak base resin offers good organic fouling protection to the downstream strong base anion resin.

5. $\overset{\downarrow}{|\text{WAC}|}-\overset{\downarrow}{|\text{SAC}|}-\text{DG}-\overset{\downarrow}{|\text{WBA}|}-\overset{\downarrow}{|\text{SBA}|}$

This scheme, whilst high on capital cost, gives overall high coflow regeneration efficiencies and therefore low unit regenerant costs because full advantage is gained from the weak function resins, and thoroughfare regeneration.

COUNTERFLOW SYSTEMS

As discussed in Chapter 7 (Column Breakthrough and 'Leakage') counterflow regeneration designs not only provide for better operational efficiencies compared with coflow systems but 'leakage' residuals are virtually eliminated such that the column effluent quality only begins to deteriorate as the resin begins to truly exhaust.

Consider the following simple fully counterflow system:

$\overset{\uparrow}{|\text{SAC}|}-\text{DG}-\overset{\uparrow}{|\text{SBA}|}$ —— service
                          --- regeneration

Given the complete regeneration of the bottom (service outlet) region of the resin bed 'leakage' residuals should be virtually nil, and indeed residuals of sodium (from the SAC) and silica (from the SBA Types 1 or 2) are typically of the order of < 0.005 mg(Na^+ or SiO_2) l^{-1} for a significant part of the run. This degree of sodium leakage present as sodium hydroxide at the anion column outlet would equate to a final conductivity of < 0.1 μS cm^{-1}. A typical counterflow quality profile is shown in Figure 8.15 where the conductivity values would not seem to support the aforementioned absolute leakage concentrations.

Two reasons are responsible for the albeit excellent quality profile being worse than theoretical. Detailed studies have shown that whilst during the early part of the cycle the cation column sodium leakage is virtually absent it is significantly greater at the anion column outlet, decreasing as the run progresses. Towards the end of the cycle the quality, as indicated by conductivity, deteriorates again due to the approach of exhaustion of either the cation bed (sodium displacement) or anion bed (hydrogencarbonate and 'silica' displacement). If anion column exhaustion *only* is controlling, the low conductivity of 'silica' and hydrogencarbonate, both eluted as their respective weak

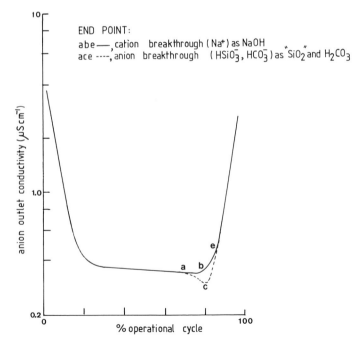

Figure 8.15 *A typical counterflow demineralized water conductivity profile*

acids, gives rise to an outlet conductivity drop prior to a steep rise upon the breakthrough of strong acid (see Figure 8.15). 'Silica' alone has virtually no conductivity which means that upon reaching conductivity 'exhaustion' on the anion column a significant amount of 'silica' may have been displaced to service.

Whilst the end of run quality profile is entirely predictable it remains to explain the source of sodium residual at the beginning of the run and it is in fact due to traces of sodium hydroxide diffusing out from within the anion resin beads after regeneration. Since prior to anion resin exhaustion the only significant demineralized water contaminant is sodium hydroxide, a properly designed hydrogen cycle strong acid cation exchange column placed downstream of the strong base anion unit will efficiently remove the sodium to produce very near theoretical water conductivity according to the reaction:

$$RH + NaOH \rightarrow RNa + H_2O \quad (8.24)$$

Just as regeneration efficiencies are improved by weak–strong functionality combinations in coflow designs so the same applies to the equivalent counterflow systems. Also, in the latter case, the lower specific gravity of specially graded weak acid and weak base resins in the hydrogen and free base forms respectively allow for thoroughfare regeneration of the weakly functional exchangers maintained above strongly functional resins in the *same vessel*. Such designs are termed *layer beds* or *stratified beds*, as denoted below:

$$\left|\frac{\uparrow WAC}{SAC\downarrow}\right| - DG - \left|\frac{\uparrow WBA}{SBA\downarrow}\right|$$

COMBINED CYCLE SINGLE STAGE DEMINERALIZATION

Mixed Bed H^+/OH^- Cycle

As the name implies, demineralization occurs through ideally simultaneous exchange of cations and anions across a uniform mixture of strong acid cation and strong base anion (Type 1) resins in the hydrogen and hydroxide forms respectively, all contained in a single vessel:

$$RH + ROH + \text{cations} + \text{anions} \rightarrow R(\text{cations}) + R(\text{anions}) + H_2O \quad (8.25)$$

Ideally a 1:1 equivalent mixture of cation and anion exchange resins may be considered as an infinite multitude of uniformly distributed 'cation–anion' pairs effecting demineralization in a single pass and the construction of such a unit is discussed briefly under Chapter 10.

Regeneration of the resins is complex and involves the following steps which are diagrammatically shown in Figure 8.16.

i) Separation: The less dense anion resin is separated from the heavier cation resin by backwashing the mixed resins, and then allowing the bed to settle. At this point the anion resin is layered above the cation resin.

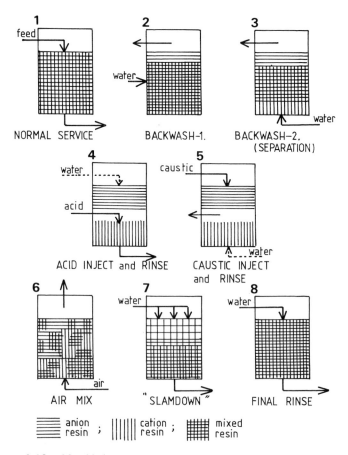

Figure 8.16 *Mixed bed regeneration sequences*

ii) Regeneration: Dilute sodium hydroxide solution of concentration about 1.25 keq m^{-3} followed by a displacement rinse are both passed downflow through the anion resin and exit the column via a collector situated at the separation interface between the two resins. The interface caustic collector now becomes a distributor to admit downflow dilute sulfuric or hydrochloric acid of concentration between 0.3–1.4 keq m^{-3} again followed by a displacement rinse both passing to drain via a bottom collecting system. The column is then drained to just above bed level and low pressure air passed upwards from the outlet collector through the entire bed to re-mix the resins. Finally the vessel is rapidly filled and the bed rinsed until the specified demineralized quality is achieved. The technology of mixed bed design has evolved over some 40–50 years and was driven by the needs of many industries for greater purity water.

Working Mixed Beds

Working mixed beds are essentially 1:1 equivalent mixes of H$^+$/OH$^-$ strongly functional resins used to demineralize raw water in a single pass and regenerated in the same vessel, termed *insitu regeneration*. A problem arises during regeneration at the separation interface region where inevitably the anion resin regenerant (sodium hydroxide solution) may contact a small region of the exhausted cation resin, whilst cation resin regenerant (dilute acid) may contact the bottommost layers of the regenerated anion resin. Despite various steps taken to minimize the degree of cross contamination there always exists the possibility of undesirable ion exchange precipitation reactions giving rise to the formation of calcium sulfate, magnesium hydroxide, and calcium carbonate which are dissipated throughout the bed after the air mix. Therefore it is usual not to guarantee a final quality better than conductivity 0.5–1.0 μS cm^{-1} and silica 0.05 mg(SiO$_2$) l^{-1}.

'Make-up' Polishing Mixed Beds

In this application a mixed bed is widely used to remove or '*polish*' the ion residuals remaining in demineralized water after prior stage-wise coflow or counterflow demineralization. The anion:cation resin volume ratio often differs from 1:1 for 'polishing' mixed beds depending upon the anticipated loading of cations (sodium) and anions ('silica'). Since no hardness cations are present, only sodium, precipitation reactions are absent and a well designed unit with clean resins can readily achieve a final quality of < 0.1 $\mu S\,cm^{-1}$ conductivity and silica < 0.02 mg(SiO$_2$) l^{-1}.

This approach of stage-wise demineralization followed by mixed bed polishing represents the long established classical process philosophy for the production of high quality make-up demineralization water for high pressure steam generation and manufacturing process waters.

Condensate Polishing Mixed Beds

The steam generation and condensing cycle is, *by itself*, not highly efficient as may be predicted by the fundamental thermodynamic theory of a reversible heat engine (Carnot Cycle) which gives:

$$\text{Thermal efficiency} = \frac{T_2 - T_1}{T_2} \qquad (8.26)$$

where T_2 is the absolute temperature (in K) of the hot source (steam) and T_1(K) the absolute temperature of the 'cold' sink (cooling water). With T_1 essentially fixed, it is readily recognizable from equation 8.26 why such great technological strides have been made in steam generator design to allow very high operating pressures and temperatures so as to maximize thermal efficiency. The pressure–temperature conditions at which an advanced power plant may operate is of the order of 10 345–22 250 kPa at superheat temperatures of about 500 °C.

The stringent design and operational demands imposed by such high pressure steam generation on the steam–water chemistry and materials of construction results in the quality of steam condensate not being adequate for direct return as boiler feedwater. Instead the condensate has to be 'polished' to remove corrosion products, ionic residuals entering with the 'make-up' water, and possible ion ingress from a condenser leak. A much simplified diagram of a power plant steam–water circuit is shown in Figure 8.17.

It is not possible to enter into great detail concerning various condensate polishing plant designs, but it is highly relevant to discuss, albeit briefly, how some ion exchange chemistry considerations impinge on current design principles. The latest designs of steam generators for power production call for a feedwater quality containing sub-microgram per litre ion concentrations of sodium, chloride, sulfate, and an absolute conductivity virtually equal to that of pure water at a specified temperature. Highly rated fossil fuel and nuclear power plant recycle condensate at a rate of approximately 3 m^3 h^{-1}

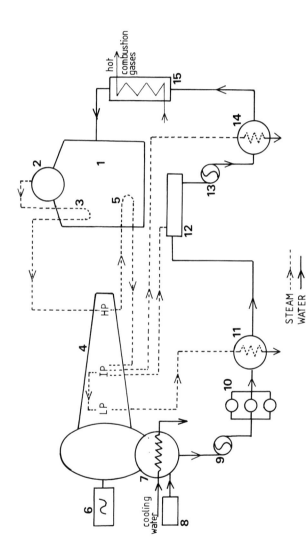

Figure 8.17 A simplified representation of a power plant steam–water circuit. Legend: 1: boiler; 2: steam–water drum; 3: superheater; 4: turbine, high (H), intermediate (I), low (L) pressure; 5: reheater; 6: generator; 7: condenser; 8: 'make-up' water treatment plant (MUP); 9: condenser extraction pump(s); 10: 'condensate polishing' water treatment plant (CPP); 11: LP heater(s); 12: deaerator; 13: boiler feed pump(s); 14: HP heater(s); 15: economizer

per megawatt, *i.e* about 6000 m³ h⁻¹ for a 2000 megawatt power station. Several mixed bed units in parallel are on-line at any given time to handle such flowrates, each operating at a linear flowrate of up to 120 m h⁻¹. Although graded gel resins offer the better regeneration efficiencies, and therefore arguably lower leakage, hydraulic constraints arising from pressure differentials and resin transfer sometimes lend preference towards macroporous products. Total bed depths of 1–1.5 m are commonly employed to handle 'clean' condensate ionic loadings of about 10 μg l⁻¹ increasing by two or three orders of magnitude under severe condenser leak conditions.

Mixed Bed H/OH Cycle: Consider leakage from the mixed bed being in the form of a neutral salt such as sodium chloride. Furthermore, assume the reasonably accurate conductivity relationship for very dilute solutions which gives:

$$0.02 \text{ meq(NaCl)} \text{ l}^{-1} \equiv 2.5 \text{ } \mu\text{S cm}^{-1}$$

or
$$1.17 \text{ mg(NaCl)} \text{ l}^{-1} \equiv 2.5 \text{ } \mu\text{S cm}^{-1}$$

From the above relationship it is readily calculated that 1 μg(NaCl) l⁻¹ made up of 0.4 μg(Na⁺) l⁻¹ and 0.6 μg(Cl⁻) l⁻¹ has a conductivity of 0.002 μS cm⁻¹. Therefore, if the conductivity of pure water is taken to be 0.044 μS cm⁻¹ at 18 °C the latter value taken as an indicator of treated condensate quality is, by itself, meaningless. Against the requirement for better than 1 μg l⁻¹ leakage of Na⁺, Cl⁻, and SO₄²⁻ the equilibrium exchange between pure water and resins partially in the Na⁺ or Cl⁻/SO₄²⁻ form cannot be ignored. It may be shown that to achieve < 1 μg l⁻¹ leakage of either Na⁺ or Cl⁻ at neutral pH on typical condensate polishing resins the degree of regeneration (conversion) to cation (H⁺) sites and anion (OH⁻) sites must be at least 61% and 12% respectively, the precise figures depending upon the values taken for the K_H^{Na} and K_{OH}^{Cl} selectivity coefficients. The fractional conversions need to be even higher under acid or alkaline conditions which could prevail if cation exchange precedes anion exchange or *vice versa* respectively. Sulfate ion leakage would ideally be expected to be low given its much greater selectivity for the anion resin over chloride, but high sulfate residuals can arise through other phenomena besides simple elution leakage:

1. Sulfate–Hydrogensulfate Hydrolysis: If sulfuric acid is used as the cation resin regenerant, under insitu regeneration a portion of anion resin at the separation interface zone becomes the hydrogensulfate form according to the reactions:

$$2\,ROH + H_2SO_4 \rightarrow R_2SO_4 + H_2O \tag{8.27}$$

$$R_2SO_4 + H_2SO_4 \xrightarrow[\text{acid}]{\text{excess}} 2\,RHSO_4 \tag{8.28}$$

After remix, and during the rinse, the prevailing neutral pH condition leads to hydrolysis of the hydrogensulfate ion which reverts to sulfate and releases sulfate ions as sulfuric acid:

$$2\,RHSO_4 \xrightarrow[\text{neutral pH}]{H_2O} R_2SO_4 + H_2SO_4 \tag{8.29}$$

The duration of this reaction is relatively short since the hydrogensulfate hydrolysis reaction is quite fast.

2. *Hydrolysis of weakly functional sites*: As a fraction of strong base sites on the anion resin degrade to weak base these sites too are able to react with regenerant acid at the interface zone:

$$RN(CH_3)_2 \xrightarrow[\text{or HCl}]{H_2SO_4} \begin{array}{l} RNH(CH_3)_2{}^+\,HSO_4{}^- \\ RNH(CH_3)_2{}^+\,\tfrac{1}{2}SO_4{}^{2-} \\ RNH(CH_3)_2{}^+\,Cl^- \end{array} \tag{8.30}$$

Hydrolysis during the final rinse and service cycle leads to the prolonged leakage of acid sulfate or chloride from the bottom of the mixed resin bed:

$$\begin{array}{l} RNH(CH_3)_2{}^+\,HSO_4{}^- \\ [RNH(CH_3)_2{}^+]_2\,SO_4{}^{2-} \\ RNH(CH_3)_2{}^+\,Cl^- \end{array} \xrightarrow{H_2O} \begin{array}{l} H_2SO_4 + RN(CH_3)_2 \\ 2\,H_2SO_4 + RN(CH_3)_2 \\ HCl + RN(CH_3)_2 \end{array} \tag{8.31}$$

3. *Kinetic Leakage*: If the anion resin is fouled in some way, for example with organics, the kinetics of exchange are impeded and detailed studies have demonstrated that the normal equilibrium selectivity of sulfate over chloride is violated leading to the preferential slip of the kinetically slower diffusing sulfate ion.

Clearly, the following considerations detract from obtaining the ultimate performance from an insitu regenerated mixed bed on condensate polishing duty:

1. Cross contamination of resin/regenerant at the separation interface.

2. Cross contamination of resin/regenerant through imperfect separation of the two resins.
3. Leakage perturbations through imperfect mixing of the two resins.

Therefore to achieve the feedwater quality required for very high pressure steam generators it is plainly advantageous to regenerate each resin physically remote from the other. An early successful compromise for insitu regeneration designs involved incorporating a small volume of *inert resin* copolymer beads of such density and size that following bed separation formed a shallow layer (approx. 15–20 cm depth) between carefully graded cation and anion resins. The purpose of the inert layer is to act as a buffer zone between the resins so minimizing regenerant cross contamination. It was soon realized that a better solution was to adopt a multivessel scheme the principle of which is shown schematically in Figure 8.18.

The treatment or service cycle occurs across a mixed bed contained in the *operator* vessel. Upon termination of the service cycle all the resin is hydraulically transferred to a *separation–regenerator vessel* for separation by backwashing. Instead of adopting insitu regeneration in this vessel the anion resin is further physically transferred to another column for separate regeneration, whilst the cation resin is separately regenerated in the separation vessel. Finally, the cation resin is transferred to the anion vessel, air mixed, rinsed, and returned to the operator.

All installed modern condensate polishing mixed bed designs adopt variations on the above described scheme, and include one or all of the following features:

1. Regeneration of cation and anion resins occurs in separate vessels.
2. An additional transfer step is often carried out to remove the unavoidable mixed resin zone at the separation interface.
3. Special regeneration practices are often employed to enhance the displacement of sodium and chloride ions from residual sites.

Mixed Bed NH_4/OH Cycle: High pressure steam generators invariably operate on an 'All Volatile Treatment (AVT)' or 'zero solids' treatment which means that any conditioning of the water–steam circuit uses chemicals which do not increase the dissolved solids of the feedwater. One such chemical is ammonia which is dosed to give a feedwater pH of 8.8–9.6 depending upon the materials of construction, and being volatile is returned with the condensate. Polishing

Water Treatment

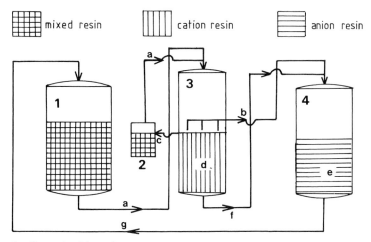

1 Operator Vessel
2 'Shuttle Tank'
3 Separator and Cation Regenerator
4 Anion Regenerator and Air Mix/Storage Vessel

Figure 8.18 *Basic sequences in condensate polishing mixed bed regeneration. Legend:*
  a) *'Exhausted' resin to Separator including interface zone from previous regeneration cycle*
  b) *Resin separation and transfer of anion resin*
  c) *Interface layer transferred to 'Shuttle Tank'*
  d) *Regeneration and rinse of cation resin*
  e) *Regeneration and rinse of anion resin*
  f) *Transfer of regenerated cation resin to anion regeneration vessel followed by air mix and final rinse*
  g) *Transfer of mixed resin to Operator*

condensate on the H/OH cycle will remove the ammonia which then has to be redosed downstream of the condensate polishing plant.

Therefore during the 1980s condensate polishing plant designs were advanced to allow the cation resin of the mixed bed to exhaust on ammonia (NH_4^+ ion) and thereafter to continue to self exchange with ammonium cation, thereby retaining the ammonia in the circuit. However, very different and adverse ion exchange equilibrium conditions apply compared with the H/OH cycle. Now, instead of neutral or near neutral exchange occurring to give only water, the combined NH_4^+/OH^- cycle exchange product is dilute ammonium hydroxide which dissociates thus:

$$NH_3 + H_2O \rightleftharpoons NH_4OH \rightleftharpoons NH_4^+ + OH^- \qquad (8.32)$$

Leakage of sodium and chloride ions is no longer the result of equilibrium with H^+ and OH^- resulting from the very weak dissociation of water, but instead with relatively significant concentrations of NH_4^+ and OH^- from the dissociation of ammonium hydroxide. Equilibrium calculations show that to achieve sub-microgram per litre residuals of Na^+ and Cl^- the fraction of sodium and chloride sites remaining on the cation and anion resin after regeneration must be less than 0.1% and 2% respectively for a system at pH 9.6 equivalent to $2.2 \text{ mg}(NH_3)\, l^{-1}$.

Clearly, resin transfers must be complete with no resin left behind, their separation perfect to avoid resin and regenerant cross-contamination and finally remixed to give a perfectly homogeneous distribution. This sets up a debate as to whether or not a mixed bed design and low ion leakage are in fact compatible ideals. Hence one major UK water treatment company has developed, and commercially operated, condensate polishing plant still based on the remote regeneration of strongly functional resins, but operated in a cation–anion–cation sequence within a single operator column, the resins being separated by robust screens. In this way the problems of resin separation and cross-contamination of resin/regenerant are avoided.

Ultrapure Water and Mixed Beds

'Ultrapure' water, whilst a term befitting to describe the ionic purity of demineralized steam condensate, has in fact passed into the language to describe the quality of water required for rinsing the etched and metallized surfaces of semiconductors and for the preparation of certain pharmaceutical formulations. Over and above the high ionic purity requirements, such treated water is also required to meet a demanding specification with regard to levels of particulates, organics, and bacteriological matter.

Table 8.4 shows the Integrated Circuit Manufacturer's Consortium suggested specification for rinse water quality in 1986. Not only is it ideally necessary to achieve sub $\mu g\, l^{-1}$ concentrations in respect of ionic constituents but demanding purity limits are set for particle counts, particle size, TOC, and bacteriological activity. Today's requirement sees the ideal level shown in Table 8.4 becoming the current target level with an even more stringent ideal level as the number of circuits per unit area required on a semiconductor wafer ('chip') is increased with advances in electronics technology.

Figure 8.19 shows a typical layout of the key operations required for

Table 8.4 *A suggested specification for electronics grade ultrapure water (Integrated Circuit Manufacturer's Consortium, 1986)*

Parameter	Ideal level	Target level	Alarm level
Resistivity (min. MΩ cm at 25 °C)	18	—	14
Copper (max. $\mu g\,l^{-1}$)	0.02	0.1	0.5
Aluminium (max. $\mu g\,l^{-1}$)	0.2	0.5	1.0
Potassium (max. $\mu g\,l^{-1}$)	0.05	0.5	4.0
Sodium (max. $\mu g\,l^{-1}$)	0.1	1.0	5.0
SiO$_2$ (total) (max. $\mu g\,l^{-1}$)	0.5	2.0	5.0
Iron (max. $\mu g\,l^{-1}$)	0.02	0.1	0.2
Zinc (max. $\mu g\,l^{-1}$)	0.02	0.1	0.5
Chromium (max. $\mu g\,l^{-1}$)	0.02	0.1	0.5
Manganese (max. $\mu g\,l^{-1}$)	0.05	0.5	1.0
Chloride (max. $\mu g\,l^{-1}$)	0.05	0.2	1.0
Nitrate (max. $\mu g\,l^{-1}$)	0.1	0.5	2.0
Phosphate (max. $\mu g\,l^{-1}$)	0.3	0.5	2.0
Sulfate (max. $\mu g\,l^{-1}$)	0.3	1.0	2.0
Particle counts (max. l^{-1})			
0.5–1 μm	50	200	1000
1–2 μm	0	10	50
TOC (max. $\mu g\,l^{-1}$)	50	100	400
Living organisms	< 1 ml^{-1}	< 5 ml^{-1}	< 10 ml^{-1}

the production of 'Ultrapure' water. The primary circuit is essentially a 'make-up' loop incorporating commonly counterflow two-stage demineralization with mixed bed or cation column polishing feeding a small storage tank suitably protected against atmospheric bacteriological ingress. A pretreatment stage may be required depending upon the quality of the raw water, for example, media filtration, organic removal, ultrafiltration (UF), and even reverse osmosis (RO).

The polishing or secondary loop usually utilizes mixed bed ion exchange and ultraviolet radiation sterilization with pre- and post-submicron membrane filtration prior to take off at the points of use. Often additional point of use membrane filtration (not shown) is employed rated at 0.2 μm or lower. Large plants may employ regenerable mixed beds and therefore adopt regeneration techniques similar to those for condensate polishing, but counterflow two-stage polishing is also viable. Alternatively, smaller demands may opt for non-regenerable cartridge mixed beds using 'semiconductor' or 'nuclear' grade resins which are supplied pre-regenerated to a very high percent conversion and low in leachable impurities.

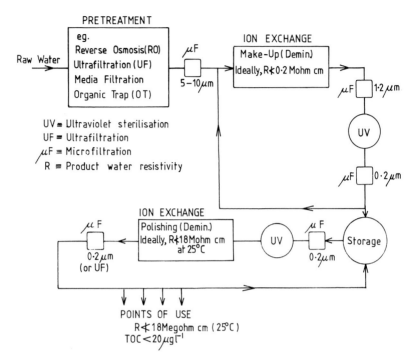

Figure 8.19 *The basic flow diagram for an ultrapure water circuit*

Stagnation of water during low or zero production demand is avoided at all costs so as not to promote bacteriological or biological contamination of either circuit. Therefore plants are designed to recirculate water in the polishing loop at least three times the maximum take off rate ($m^3 h^{-1}$) at velocities of 2–3 m s^{-1} in delivery pipework free of 'dead legs'.

DESALINATION BY ION EXCHANGE

Demineralization by ion exchange usually involves chemical regeneration of the resins with strong acid or alkali solutions. Weaker electrolyte regenerants are sometimes employed such as solutions of carbon dioxide, ammonia, or lime as demonstrated by various novel processes such as *Desal* and *Carix* for the partial demineralization, or desalination, of brackish waters. The increased dissociation of the *salt* forms of weakly functional cation and anion exchange resins at increased temperatures is the basis of the *Sirotherm* process which uses alternate hot and cold cycling of special mixed-functional resins to effect partial demineralization:

$$\text{RN(CH}_3)_2 + \text{RCO}_2\text{H} + \text{H}_2\text{O} + \text{NaCl} \underset{\underset{\text{Regeneration}}{(80\,°C)}}{\overset{\overset{\text{Service}}{(20\,°C)}}{\rightleftharpoons}}$$
$$\text{RCO}_2\text{Na} + \text{RNH(CH}_3)_2{}^+\text{Cl}^- + \text{H}_2\text{O} \quad (8.33)$$

The Sirotherm Process is an example of non-chemical regeneration.

The most widely known, and applied, ion exchange process which does not rely upon chemicals for resin regeneration is *electrodialysis*, which uses ion exchangers in the form of heterogeneous or homogeneous membranes (see Chapter 2, 'Special Ion Exchange Materials'), for the partial demineralization of brackish waters containing around 2–12 kg m^{-3} dissolved solids. The main features of the operation are shown in Figure 8.20. The feed is passed through a series of 'stacks' each of which contains a large number of closely separated ion exchange membranes of strong acid/strong base type arranged in alternate fashion. The cationic membranes B allow only the passage of cations, while anion exchange membranes A are selective to the passage of anions. The applied DC potential (1–2 volts per pair of

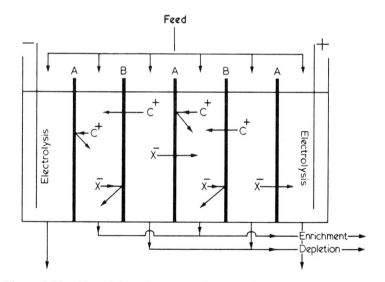

Figure 8.20 *Electrodialysis. A = anion exchange membrane*
B = cation exchange membrane
(Reproduced from R. W. Grimshaw and C. E. Harland, 'Ion-exchange: Introduction to Theory and Practice', The Chemical Society, London, 1975)

membranes) and the polarity cause a migration of ions to take place in the direction indicated in the diagram, which causes enrichment and depletion of electrolyte in alternate stack compartments. Electrolysis occurs at the terminal electrodes and therefore it is important to be able to deal effectively with corrosive gases which are generated, for example chlorine.

Demineralizing brackish waters to below about $0.3–0.4\,\text{kg m}^{-3}$ is economically unattractive by these means, since the increasing electrical resistance in the dilute compartments would demand an increased power requirement, but otherwise the technique is competitive with other methods. Demineralization to a low level can be achieved by packing the diluant compartment with mixed strongly functional resins in the hydrogen and hydroxide form. The conducting path offered by the resins and their autoregeneration through the dissociation of water enables low diluant conductivity to be achieved. This is the principal of *continuous electrodeionization* (CDI) which for a low conductivity feedwater $(50\,\mu\text{S cm}^{-1})$ together with membrane filtration claims to produce a final water of Ultrapure quality.

WASTE EFFLUENT TREATMENT BY ION EXCHANGE

An increasing awareness of environmental legislation has encouraged the practice of more efficient liquid waste management and to the reuse of process water. The present areas of application are mainly in the treatment of obnoxious effluents from the metal fabrication industries and radioactive liquid streams where economic benefits result from recovering reagents and recovering clean water. The recovery of water from treated sewage effluents is of major interest and pilot-plant scale ion exchange studies have been undertaken in this field.

Chemical Waste Streams

The first large scale application of ion exchange to effluent treatment was in the recovery of water, ammonia, and basic copper sulfate from the waste streams encountered in the cuprammonium rayon process. Originally a phenolic type condensation resin was employed, but more recently carboxylic acid acrylic-based exchangers have been introduced. A similar process exists for zinc recovery from the spinning acids of viscose rayon plants, except that in this operation a sulfonic acid resin is employed.

Ion exchange methods are established for treating various rinse streams arising from metal finishing processes such as plating and anodizing. Thus in chromic acid anodizing of aluminium, the demineralization reactions may be represented thus:

Cation Column

$$6\,RH + (M^{III})_2(Cr^{VI}O_4)_3 \xrightarrow{\text{loading}} 2\,R_3M^{III} + 3\underset{\text{chromic acid}}{Cr^{VI}O_3.H_2O} \quad (8.34)$$

$$2\,R_3M^{III} + 3\,H_2SO_4 \xrightarrow{\text{regeneration}} 6\,RH + (M^{III})_2(SO_4)_3 \quad (8.35)$$

where ions M^{III} are tervalent aluminium, chromium, and ferric iron.

The first regeneration of the strong base anion exchange resin with a near stoichiometric quantity of sodium hydroxide converts the loaded dichromate form of the resin to the chromate form whereafter it is able to efficiently take up chromic acid again. The anion column effluent of sodium chromate may be cation exchanged across a strong acid resin in the hydrogen form to recover chromic acid:

Anion Column

$$R_2Cr^{VI}O_4 + Cr^{VI}O_3 \xrightarrow{\text{loading}} R_2Cr_2^{VI}O_7 \quad (8.36)$$

$$R_2Cr_2^{VI}O_7 + 2\,NaOH \xrightarrow{\text{regeneration}} R_2Cr^{VI}O_4 + Na_2Cr^{VI}O_4 + H_2O \quad (8.37)$$

Acid Recovery

$$2\,RH + Na_2Cr^{VI}O_4 \rightarrow 2\,RNa + Cr^{VI}O_3.H_2O \quad (8.38)$$

Clearly, ion exchange methods can overcome what would otherwise be a difficult waste disposal problem, and the bonuses of recovering reagents and process water are valuable economically. A diagrammatic representation of the ion exchange route is shown in Figure 8.21.

In electroplating operations the rinse stream is complicated by the presence of complex anions and organic additives. Although the recovery of reagents is more difficult in these cases, demineralization is feasible by ion exchange techniques involving upstream sorption processes on activated carbon to remove detergents and organic conditioners followed by conventional two-stage demineralization to recover the rinse water. The concentrated resin regeneration wastes are treated chemically for ultimate safe disposal as a dewatered solid.

Ion exchange treatments have successfully been applied to the effluent streams arising from paper manufacture, photographic processing, chemical leaching, zinc smelting, and metal pickling. The

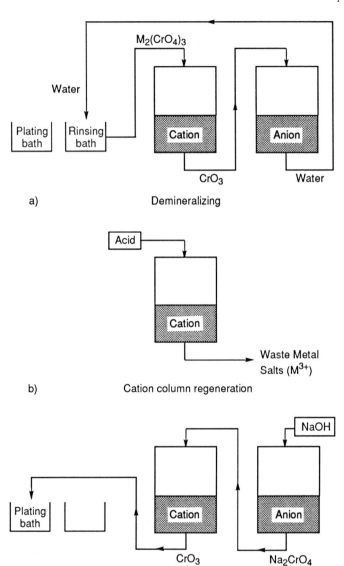

Figure 8.21 *Schematic process diagram for chromic acid recovery and water reuse in the aluminium anodizing process*
(Adapted from R. W. Grimshaw and C. E. Harland, 'Ion-exchange: Introduction to Theory and Practice', The Chemical Society, London, 1975)

latter furnishes a particularly interesting example of cation removal as a complex anion. Concentrated hydrochloric acid used in the steel galvanizing process becomes contaminated with ferric (Fe^{III}) iron and zinc (Zn^{II}) in the form of their chloro-complex anions, $Fe^{III}Cl_4^-$ and $Zn^{II}Cl_4^{2-}$. Strong base anion exchange resins in the chloride form readily take up the chloro-complex ions thereby rejuvenating the hydrochloric acid:

$$RCl + Fe^{III}Cl_4^- \rightarrow RFe^{III}Cl_4 + Cl^- \qquad (8.39)$$

$$2\,RCl + Zn^{II}Cl_4^{2-} \rightarrow R_2Zn^{II}Cl_4 + 2\,Cl^- \qquad (8.40)$$

Regeneration is effected with water which through hydrolysis decomposes the anionic complex ions to give a rapid elution of the cations, firstly Fe^{III} and then Zn^{II}, which are recovered (see also Box 5.2). The severe osmotic cycling between treatment and regeneration stages dictate the use of macroporous resins.

Often the removal of low levels of toxic ionic pollutants from industrial waste streams on conventional ion exchange resins is made difficult by their non-selectivity against a high concentration of competing ions, *e.g.* Na^+, SO_4^{2-}, Cl^-, H^+, OH^-. Therefore in the light of current awareness relating to the aqueous environment a renewed interest is being taken in a possible more widespread use of chelating resins and powdered sorbents in this area.

Radioactive Streams

The most difficult waste disposal problems arise from the effluents produced during the processing of spent nuclear fuel where liquid–liquid extraction techniques are used to recover fissile and fertile material from the various fission products.

Radioactive effluents are classified as being of high, medium, or low activity and ion exchange treatments may be applied to all three types. Medium and low level effluents are often treated chemically for the removal of activity. Alternatively, they may be discharged to a ground site which contains an adequate depth of ion exchanging soil – and suitable geological and hydrological properties – to retain the activity safely. Increased restrictions on the discharge of low and medium level liquid radioactive wastes have re-established the role of synthetic aluminosilicate exchangers which are relatively cheap and are more stable towards high temperature and radiation breakdown than resin forms. Furthermore, they exhibit high affinities towards certain long half-life species commonly constituting radioactive wastes, *e.g.* ^{137}Cs, and large volumes of solution can be treated before breakthrough occurs.

Highly active effluents present a more serious problem since they must be stored virtually indefinitely in leakproof and shielded underground bunkers. The current practice is to concentrate the activity and reduce its volume by evaporation, but this is only a partial solution. Species with long half-lives, for example ^{137}Cs and ^{90}Sr, can be retained on synthetic and naturally occurring aluminosilicates which thus concentrate the dangerous isotopes in a solid waste of relatively small volume. The fixation of radioactive species on aluminosilicates by calcination, or by forming non-leachable glasses are two methods which may be adopted for permanently containing the activity.

Ion exchange on resins is an important feature of nuclear power generation, not only for 'make-up' water and steam generator condensate polishing circuits but also, depending upon the reactor design, for chemical control of coolant/moderator systems for the 'boiling', 'pressurized', and 'heavy' water reactors commonly designated BWR, PWR, and HWR respectively. All designs, including gas cooled reactors, incorporate ion exchange circuits for decontaminating active effluents.

Gas Cooled Reactors

Gas cooled reactors use carbon dioxide under pressure as a recirculating heat transfer medium (coolant) between the hot nuclear reactor core and water in a secondary circuit in order to raise steam and electrical power in an otherwise conventional high pressure steam generator/turbine/condenser loop. The role played by ion exchange is denoted by systems A–D in Figure 8.22.

The water 'make-up' (A) and condensate polishing (B) systems are similar in design to those installed in fossil fuel power plants. Spent nuclear fuel is stored in a cooling pond under very pure demineralized water which is conditioned to a pH suited to the non-corrosivity of the fuel rod cladding. With time the pond water becomes contaminated with particulate matter and dissolved radioactive species, such as ^{137}Cs. Decontamination of the pond water is achieved by filtration and demineralization across a combination of cation exchange followed by mixed beds (C). The magnesium alloy cladding of Magnox reactor fuel requires pond storage at quite high pH such that cation exchange on a methylenesulfonate phenolic condensation resin is often used

Water Treatment

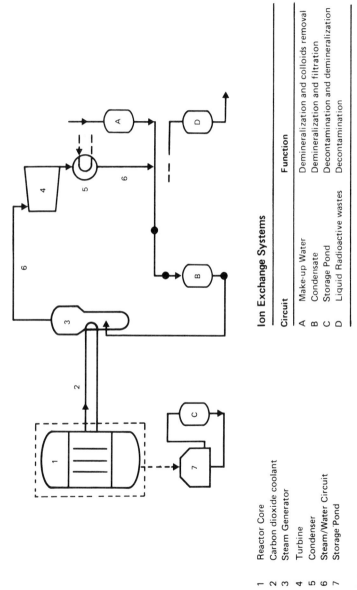

Figure 8.22 *Schematic diagram of a nuclear gas cooled reactor showing the location of ion exchange treatments* [From 'Amberlite Ion Exchange Resins in Nuclear Power Technology', Rohm and Haas (European Region)]

which shows a high affinity towards caesium even if present in a high background concentration of sodium alkalinity (see Table 2.1). Ion exchange resins used for nuclear decontamination processes are of extremely high purity (Nuclear Grade) to minimize 'leachables' that could become activated and add to the level of activity in a circuit. Also the resins are supplied in a virtually 100% regenerated state in the desired ionic form since resins exhausted on active species are rarely regenerated but sluiced from the vessel to an active waste containment facility.

Waters used to transfer active exhausted resins, laboratory active effluents, and laundry wastes all require decontamination before discharge which uses ion exchange on separate or mixed beds (D).

Water Cooled Reactors

There are several operating designs of nuclear reactor which use water both as a neutron flux moderator and as a coolant to transfer heat directly (Boiling Water Reactor) or indirectly (Pressurized Water Reactor) to a steam generator. The Pressurized Water Reactor (PWR) shown simplistically in Figure 8.23 is particularly interesting since ion exchange is extensively used to control the system chemistry. Filtration and full conventional condensate polishing is employed to demineralize the condensate from the secondary water circuit feeding the steam generator, which may also contain active species arising from leakage between the primary (coolant) and secondary circuits.

Boron in the form of boric acid may be added to the primary coolant which also serves as a further moderator and together with added lithium hydroxide controls the corrosivity of the primary coolant towards the system components. The hydroxide of the lithium seven isotope (^{7}Li) is used for pH control of the primary coolant since any appreciable amounts of ^{6}Li isotope produces the gaseous highly radioactive hydrogen isotope tritium (^{3}H) by neutron capture:

$$^6_3\text{Li} + ^1_0\text{n} \rightarrow ^4_2\text{He} + ^3_1\text{H} \tag{8.41}$$

Lithium will also accumulate in the circuit due to irradiation of the ^{10}B isotope which is naturally present in the boric acid:

$$^{10}_5\text{B} + ^1_0\text{n} \rightarrow ^7_3\text{Li} + ^4_2\text{He} \tag{8.42}$$

The activity of the primary coolant may be controlled by side-stream demineralization (decontamination) across mixed beds (C) of

Water Treatment

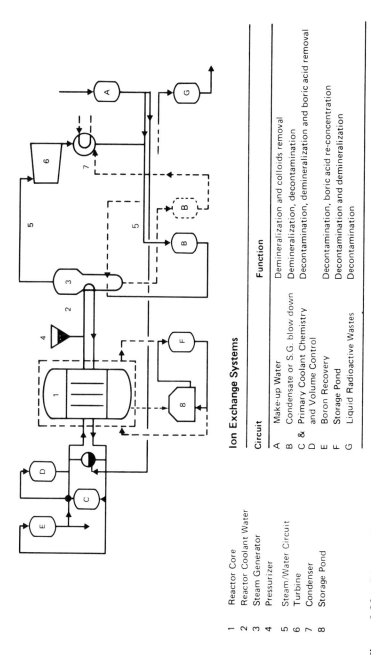

Figure 8.23 *Schematic diagram of a pressurized water reactor showing the location of ion exchange treatments* [From 'Amberlite Ion Exchange Resins In Nuclear Power Technology', Rohm and Haas (European Region)]

cation and anion exchanger resins in the lithium (^{7}Li) and borate form respectively, thus maintaining the desired levels of lithium hydroxide and borate in the coolant. The excess build-up of lithium in the coolant is removed by side-stream treatment across a cation exchange bed (D) in the hydrogen form situated downstream of the decontamination units. Boron levels in the coolant may be controlled by a partial bleed off which is either discharged or the boron recovered as borate across a strong base anion exchange resin (E).

FURTHER READING

'Ion Exchangers', ed. K. Dorfner, Walter de Gruyter, Berlin and New York, 1991.

G. Solt and C. Shirley, 'An Engineer's Guide to Water Treatment', Avebury Technical, Aldershot, 1991.

S. Applebaum, 'Demineralization By Ion Exchange', Academic Press, New York and London, 1968.

T. V. Arden, 'Water Purification by Ion Exchange', Butterworths, London, 1968.

'Ion Exchange In The Process Industries', Society of Chemical Industry, London, 1970, Conf. Proc. (Cambridge).

'The Theory and Practice of Ion Exchange', ed. M. Streat, Society of Chemical Industry, London, 1976, Conf. Proc. (Cambridge).

'Ion Exchange Technology', ed. D. Naden and M. Streat, Ellis Horwood, Chichester, 1984.

'Ion Exchange For Industry: Development and Use', ed. M. Streat, Ellis Horwood, Chichester, 1988.

'Ion Exchange Advances', ed. M. J. Slater, Elsevier Applied Science, London and New York, 1992.

J. H. Smith, 'Modern Countercurrent Ion Exchange Plants and The Hipol Process', *Chem. Ind. (London)*, 1980, No. 18, 718.

T. A. Peploe, 'The Tripol Process–A New Approach to Condensate Polishing', *Chem. Ind. (London)*, 1980, No. 18, 724.

J. R. Emmett, 'Ion Exchange Fundamentals Applied to Condensate Polishing', *Chem. Ind. (London)*, 1980, No. 18, 730.

T. V. Arden and T. Hall, 'Nitrate Removal from Drinking Water – A

Technical and Economic Review', Water Resource Centre Report 856-S, 1989.

B. T. Croll, 'The Removal of Nitrate from Water using Ion Exchange', in 'Ion Exchange Processes: Advances and Applications', ed. A. Dyer, M. J. Hudson, and P. A. Williams, Special Publication No. 122, Royal Society of Chemistry, Cambridge, 1993, p. 141–158.

M. A. Sadler, 'Developments in the Production and Control of Ultrapure Water', *ibid.*, p. 15–28.

J. Lehto, 'Ion Exchange in the Nuclear Power Industry', *ibid.*, p. 39–53.

Chapter 9

Non-water Treatment Practices

CARBOHYDRATE REFINING

By non-water treatment is usually meant the treatment of an aqueous solution in order to purify the particular solute rather than the solvent, *i.e.* water.

Commercial ion exchange processes in carbohydrate treatments are concerned with the purification of juices and syrups from cane sugar, beet sugar, and corn starch hydrolysates. As with water treatment, the main operations are softening (decalcification), demineralizing (de-ashing), and decolorizing (removal of organic colour bodies), all of which improve the yield and quality of the final recrystallized sugar or concentrated syrup.

Strong acid resins in the hydrogen form catalyse the inversion of sucrose ($C_{12}H_{22}O_{11}$) to give invert sugar: an equimolar mixture of fructose and glucose which rotates plane polarized light in a direction opposite to that of pure sucrose. This in itself is a classical application of industrial catalysis by ion exchangers, but if inversion is to be avoided in two-stage demineralization, strong base anion exchange may precede strong cation exchange (reverse demineralization) thereby avoiding the formation of an intermediate acidic liquor. Alternatively a cool dilute extract ('thin juice') is passed at a high flowrate through the leading strong acid resin to reduce residence time and minimize the degree of inversion. In another option, demineralization may be achieved by mixed bed ion exchange using a weakly acidic cation exchanger and strong base anion exchange resin.

Usually raw sugar processing gives rise to fairly viscous juices and syrups at temperatures between 70–90 °C. Therefore macroporous ion exchange resins are often selected for sugar extract treatments operated as deep beds at fairly low specific flows ($m^3 \, h^{-1} \, m^{-3}{}_R$) because of the slower kinetics compared with water treatment. A most interesting application is afforded by the *Quentin Process* which is based upon the discovery in 1955 that the presence of sodium and potassium ions in beet sugar crystallization mother liquor increases the amount of sugar remaining in the discard molasses after evaporation and crystallization, thus reducing sugar yield. However the sugar loss to molasses is significantly reduced if the monovalent sodium and potassium ions are firstly exchanged for divalent magnesium across a macroporous cation exchanger in the magnesium form. Regeneration of the resin is achieved using a concentrated solution of magnesium chloride. Thus here is an example of what is plainly an unfavourable equilibrium and low efficiency process being outweighed by the economic gains in improving sugar yield by a few percent.

Recent years have seen a rapid growth in the production of dextrose [(+)-glucose, dextrorotatory] and fructose from high fructose corn syrups (HFCS). An approximately 1:1 dextrose:fructose mixture is obtained by the enzyme catalysed isomerization of dextrose obtained from corn hydrolysate, which is further purified by two-stage demineralizing (de-ashing), colour body removal across a macroporous weak base resin, and mixed bed polishing. Enriched frustose syrups required by the soft drinks and food industry as a low calorie sweetener may be obtained by a chromatographic separation process known as *ligand exchange*. The hot dextrose–fructose mixture is passed slowly down a deep column of finely sized cation exchange resin in the swollen calcium form, which shows a preferred affinity for fructose over dextrose. The sorption mechanisism is believed to be one of ligand exchange between one or several of the hydroxyl groups of the

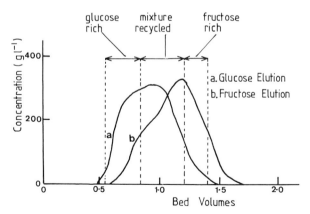

Figure 9.1 *Separation of glucose and fructose by ligand exchange across a strongly acidic cation exchange resin in the Ca^{2+} form*
(Reproduced by permission from D. Hervé, 'Ion Exchange In The Sugar Industry Parts 1 and 2', *Process Biochem.*, 1974, **9**, p. 14; p. 31)

sugar molecules with hydration water molecules of the resin counterion, rather than simple ion exclusion (see 'Ion Separation', later). Elution with demineralized water allows dextrose and fructose to be collected separately with the mixed band being recycled. A similar process may be employed to separate glucose and fructose from acid inverted sucrose (Figure 9.1).

BOX 9.1   Decolorizing and Demineralizing of a Sugar Solution

Processed crude sugar solutions may contain colouring matter and other ionic materials. The coloured ions are usually anions and may be removed by an anionic exchanger; other ions are removed by a mixture of anionic and cationic exchangers. A strong cation exchanger in the hydrogen form will cause partial inversion of the sugar; thus a weak cation exchanger should be used, coupled with a strong anion exchanger in the hydroxide form to make a mixed bed. The sugar molecules are not ion exchanged, and apart from some initial dilution, pass straight through the resin bead.

This method of demineralizing organic non-polar compounds is a general application.

Requirements

12 mm diameter column (approx.)
25 ml measuring cylinder
Conductivity meter
Refractometer
20 ml of SBA resin (Cl form)
10 ml of WAC resin (H form)
Absorptiometer or colour comparator (not essential)
1 N sodium hydroxide
500 ml of 20% w/v sugar solution (use white granulated sugar, and add brown sugar to give a straw colour and a conductivity of 100 to 200 micromhos cm^{-1})

Procedure

1. Tap down the resins under water in the measuring cylinder.
2. Put 20 ml of SBA resin (Cl form) into the column and regenerate with 40 ml of 1 N NaOH at 1 BV/10 minutes.
3. When the alkali has drained to bed level, rinse with 200 ml of deionized water at 1 BV/4 minutes.
4. Transfer the resin to a beaker and add 10 ml of WAC resin (H form). Mix the resins and return the mixture to the column. Run deionized water through the column at about 1 BV/2 minutes until the eluate has a low conductivity, *e.g.* 50 micromhos.
5. Run the sugar solution through the column at 1 BV/10 minutes. Assess

> colour removal by an absorptiometer or visually. Follow demineralizing with the conductivity meter, and sugar concentration with the refractometer. If sufficient sugar solution is run through, the colour and/or conductivity of the eluate will increase as the column becomes fully loaded.
>
> TIME REQUIRED – 3 hours

CATALYSIS

Acids (hydrogen ion) and bases (hydroxide ion) act as homogeneous catalysts for many important organic chemical reactions in solution. These include esterification, ester hydrolysis (see Box 9.2), hydration of alkenes, dehydration of alcohols, and condensation reactions.

Strongly acidic or basic ion exchange resins in the hydrogen and hydroxide forms respectively catalyse the types of reaction listed above. The kinetic mechanism is one of diffusion of the chemical reactants into the interior of the exchanger where the reaction is promoted. Although basically a heterogeneous system, the reaction is best described as being homogeneous within the gel structure, or pore volume in the case of macroporous resins. Ion exchange resins are attractive as catalysts because they are readily reclaimed by a simple filtration step which quenches the reaction rapidly. The product is catalyst-free, and harmful or competing side reactions are often less apparent than in conventional homogeneous catalysis. Ion exchangers may often exhibit selectivity towards the type of reaction that they promote, thereby improving the purity and possibly the yield of product.

Mention has been made previously of the use of acid form cation exchangers to bring about optical inversion of saccharides. Another important application in the petrochemical industry is the use of hydrogen form macroporous cation exchange resins as a catalyst in the synthesis of various alkoxyalkanes (ethers) by the electrophilic addition of alcohols across carbon–carbon double bonds of alkenes:

$$\mathrm{>\!C\!=\!C\!<} + ROH \underset{100\,°C}{\overset{\text{Resin (H}^+)}{\rightleftharpoons}} \begin{array}{c} |\ \ | \\ -\mathrm{C-CH} \\ |\ \ | \\ \mathrm{O} \\ | \\ \mathrm{R} \end{array} \qquad (9.1)$$

Methanol reacts with 2-methylpropene to give 1,1-dimethylethoxymethane (methyl t-butyl ether) or MTBE, which is an important additive to petroleum fractions for increasing the octane rating:

$$H_2C=C{\underset{CH_3}{\overset{CH_3}{\diagdown}}} + CH_3OH \underset{100\ ^\circ C}{\overset{Resin\ (H^+)}{\rightleftharpoons}} H_3C-\underset{CH_3}{\overset{CH_3}{\underset{|}{\overset{|}{C}}}}-O-CH_3 \quad (9.2)$$

2-methylpropene   methanol   methyl t-butyl ether

Resins may also act as a substrate for metallic hydrogenation catalysts such as platinum, palladium, nickel, *etc.*, but for high temperature petrochemical cracking and reforming reactions the thermally more stable zeolite products dominate this particular use.

BOX 9.2   Catalysis – Hydrolysis of Ethyl Acetate by Means of a Cation Resin

This is a particularly interesting experiment for advanced students. It may be carried out at various temperatures to show the effect of temperature. The velocity constant of reaction and energy of activation may also be determined. The reaction is followed by titrating the free acidity at timed intervals.

Requirements

Stirrer
Water bath
Stoppered 250 ml flask and reflux condenser for elevated temperature work
Thermometer
4 g (approx.) of dry resin per experiment
Dry SAC resin (H form)
2 ml of ethyl acetate per experiment
Timer
0.1 N NaOH
Phenolphthalein indicator

Procedure

Place the flask in the water bath at the required temperature, say 25 °C, and add 60 ml of deionized water. Add 4 g of dried H form resin and start the stirrer. Leave stirring for at least 15 minutes or until the water temperature equals the bath temperature.

Remove the stopper and add 2 ml of ethyl acetate by pipette. Stop the stirrer and remove a 2 ml aliquot from the flask by pipette. A stop watch is started when the addition is made, and the time and temperature taken when the first sample is removed. The stopper is replaced rapidly and the stirrer restarted. 2 ml aliquots are removed at the following times: 1, 10, 20, 30, 40, 60, and 80 minutes. The experiment is then allowed to continue undisturbed for at least 24 hours when a final sample is removed.

The progress of the experiment is followed by adding the aliquot to a flask containing about 50 ml of cool mixed bed deionized or distilled water. The acidity is titrated at once using phenolphthalein indicator and standardized 0.1 N sodium hydroxide.

Plot graphs of titre against time in minutes, and temperature against time. If time $= t$, titre at time $t = T_t$, titre at 24 hours $= T_\infty$. Plot a graph of $\log_{10}(T_\infty - T_t)$ against time.

$$\log_{10}[CH_3COOEt] = -\frac{kt}{2.303} + \text{constant}$$

$$\text{slope} = -\frac{k}{2.303}$$

Hence k, the velocity constant, can be found. From the Arrhenius relationship $k = AE/RT$ it can be shown that:

$$\log_{10}\left(\frac{k_1}{k_2}\right) = \frac{E}{R}\left(\frac{1}{T_2} - \frac{1}{T_1}\right)$$

where

$E = $ energy of activation

$R = $ gas constant

T_2 and $T_1 = $ temperature in K

Hence, since all these quantities are known, E the energy of activation can be found.

Typical results from this experiment are:

1. Velocity constant of reaction $= 2.99 \times 10^{-3}$/minute at 23.8 °C.
2. Velocity constant of reaction $= 10.22 \times 10^{-3}$/minute at 38.0 °C.
3. Energy of activation $= 26$ kJoules mole^{-1}

N.B. The inversion of a sucrose solution may be similarly investigated.

TIME REQUIRED – 2 hours (plus overnight to obtain final reading)

METATHESIS

Metathesis or ion substitution reactions using ion exchange resins could, in principle, be as widespread as the possible pairing of ions of the same charge type. In analytical chemistry the possible applications are endless, and some thought should always be given to the possible role of ion exchange to eliminate interfering ions during wet chemical analysis procedures. In essence, all ion exchange reactions are substitution or metathesis reactions but a distinction is made between a substitution which leaves the total concentration unchanged and those processes which remove ions to produce water or pure solvent, for example, the second stage of two-stage demineralization.

One well cited example is the removal of carbonate impurity from standardized sodium hyroxide solution across a column of strong base anion exchange resin in the hydroxide form thereby restoring the initial standard concentration, namely:

$$\underset{(x)\ \text{eq}\,l^{-1}}{OH^-} + \underset{(y)\ \text{eq}\,l^{-1}}{CO_3^{2-}} \xrightarrow{\text{Resin (OH)}} \underset{(x+y)\ \text{eq}\,l^{-1}}{OH^-} \qquad (9.3)$$

where

$$(x + y) = \text{Standard Concentration (eq}\,l^{-1})$$

The determination of total cations by titrating the acidity liberated by passage through a strong acid cation exchange resin in the hydrogen form (*EMA* – see Chapter 8, 'Water Analysis') is a simple example of metathesis. An unknown concentration of any cation could be determined in this manner and likewise an anion using a hydroxide form anion exchange resin providing that precipitation or gas evolution cannot occur. The disolving of precipitates by equilibration with ion exchange resins is a striking example of ion exchange metathesis (see Box 9.3).

Industrially, metathesis reactions may be used to prepare a stable silica sol by passing sodium silicate solution down a column of strongly acidic cation exchange resin in the hydrogen form (Box 9.3). Modified water soluble ethylenediaminetetraacetic acid is commercially produced by conversion of the sodium salt across the hydrogen form of strong acid resin. The same principle applies to vitamin B_5 production for converting its sodium salt form to the acid form by counterflow ion exchange across a strong acid resin.

BOX 9.3 Metathesis Reactions

Preparation of a Silica Sol Using a Cation Resin

Silica sol may be conveniently prepared from sodium silicate using a strong resin in the hydrogen form.

Requirements

25 ml of SAC resin
Glass column, 12 mm diameter (approx.)
Solution of sodium silicate containing 3% SiO_2
2 N and 5 N HCl

Procedure

Regenerate the resin with 1 BV of 2 N HCl, then 1 BV of 5 N HCl, at a rate of 1 BV/15 minutes. Rinse with 7 BV of deionized water at a flow rate of 1 BV/5 minutes. Pass 2 BV of the sodium silicate through the resin at 1 BV/15 minutes. The resulting sol will be almost clear and will remain so for a fairly long time.

TIME REQUIRED – 1 hour 45 minutes

Redissolving of a Precipitate

This is a particularly striking example as it is coloured and, although the technique may be applied to many insolubles, this reaction is rapid and is therefore suitable for lecture demonstration.

A freshly precipitated lead iodide is reacted with a cation resin in the sodium form. Exchange takes place between the lead and sodium ions and the precipitate is redissolved.

Requirements

10 ml of SAC resin (Na form)
Boiling tube
4 ml of 1% $Pb(NO_3)_2$ solution
4 ml of 1% KI solution

Procedure

Put 15 ml of deionized water into the boiling tube, add 4 ml of $Pb(NO_3)_2$ and 4 ml of the KI solution. Yellow lead iodide is precipitated. Add 10 ml of the resin and shake until the yellow precipitate disappears.

Similarly, calcium carbonate, silver chromate, chromium hydroxide, and barium sulfate may be redissolved but very extended shaking times are involved.

TIME REQUIRED – 15 minutes

RECOVERY PROCESSES

These processes use ion exchangers to retain ionic components which are then recovered as a concentrate during the subsequent elution or displacement step. In all cases the liquors to be treated are *dilute* and contain species of either high commercial value or which present particularly difficult disposal problems; sometimes both these considerations might apply. Therefore ion exchange is well placed as a processing technology as demonstrated by its widespread use in hydrometallurgy and pharmaceutical operations.

Metal Recovery from Dilute Leach Liquors

Table 9.1 lists some metals which may be recovered and purified commercially by ion exchange whether from process waste streams or primary ore treatments. Because of the increasing demand on the Worlds's natural mineral resources there is an important need for efficient, and economic, methods of treating low grade ores, waste mine waters, mineral dumps, and discarded tailings.

Ion exchange practices are widely applied commercially in South Africa and North America for the recovery and concentration of uranium oxide (U_3O_8) where, since 1945, low grade uranium ores have been treated in order to meet the needs of the World's Atomic Energy programmes. The existing 'rich' deposits assay at typically about 0.1–0.2% U_3O_8, but sources reporting as low as 0.005% may be utilized especially if uranium is a by-product of other operations such as gold recovery or phosphoric acid production from phosphate rock. Sulfuric acid leaching under oxidizing conditions is the usual method for treating uranium ores to yield a liquor containing about 0.1–1 g U_3O_8 per litre. For such low concentrations normal extraction techniques would be difficult and costly unless the liquor is firstly concentrated. The use of strong base or weak base resins for this

Table 9.1 *Metals recovered and purified commercially by ion exchange*

Uranium	Platinum metals
Thorium	Copper ⎫
Rare Earths	Zinc ⎬ Rayon Plants
Trans-uranium elements	Rhenium
Transition metals	Molybdenum
Gold ⎫	Cobalt
Silver ⎬ Plating Wastes	Nickel
Chromium ⎭	

purpose is the largest single application of ion exchange outside the field of water treatment.

The sorption of uranium from acid sulfate leach liquors by strong base anion exchange resins is unusual since complexes of the type $[UO_2(SO_4)_n]^{2-2n}$ may be sorbed by both ion exchange and addition mechanisms. High concentrations of other species are present in the leach solution due to dissolution of pyritic and siliceous components of the ore, but, apart from iron, they do not interfere with the sorption of uranium as a complex anion. Iron(III) also forms an anionic sulfate complex, but is only weakly held by the resin and is displaced ahead of the uranium. The sorption of uranium may be represented by equations of the type:

$$2\,R_2SO_4 + UO_2(SO_4)_3^{4-} \rightarrow R_4UO_2(SO_4)_3 + 2\,SO_4^{2-} \quad (9.4)$$

$$R_2SO_4 + UO_2(SO_4)_2^{2-} \rightarrow R_2UO_2(SO_4)_2 + SO_4^{2-} \quad (9.5)$$

Elution of the resin is usually carried out with sulfuric acid ($2\,\text{keq m}^{-3}$) although acid chloride and acid nitrate eluants are also used. Sulfuric acid is also a preferred eluant when ion exchange is used upstream of a liquid–liquid extraction recovery enhancement stage. The early installations employed a batch process using columns of anion exchange resin on a 'merry go round' system. Briefly, a clarified leach liquor of pH between 1.0 and 2.0 is passed downflow through two anion exchange columns in series. On breakthrough of uranium from the second column, the leading column is taken off stream and rinsed with dilute acid prior to eluting the uranium with the appropriate eluant. A fresh eluted column then becomes the second column in the loading sequence. The complete sequence of ion exchange operations showing the split-elution technique is shown in Figure 9.2. This type of elution practice is valuable in ion exchange operations since it is economical in the consumption of regenerants and so maintains a concentrated eluate. The final steps in the process are the precipitation of the diuranate salt from the concentrated eluate, which is followed by calcination of the precipitate to give a product analysing to over 90% uranium oxide (U_3O_8).

Combined ion exchange and liquid–liquid extraction processes produce an oxide which is pure enough to be a direct source of nuclear grade metal. A later development (Porter–Arden) involved transferring resins between loading and elution stations which simulated a more efficient countercurrent operation but was not truly continuous. Fixed bed designs *must* employ a clarified feed if not to

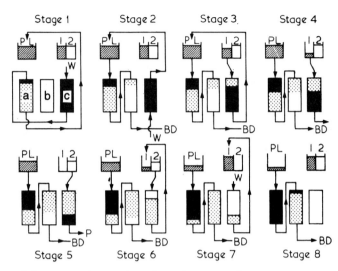

Figure 9.2 *Process diagram for a three column ion exchange uranium recovery plant.*
Legend:
Resin loading:
Stages 2 to 8   pregnant feed flow through columns (a) and (b)
Resin washing and elution:
Stage 1   displacement of pregnant feed from (c) to (a)
            displacement of feed solution from (a) to pregnant feed tank
Stage 2   water wash of column (c)
Stage 3   first eluant to (c) to displace water
Stage 4   first eluant to (c) to eluant make-up
Stage 5   second eluant to (c) peak uranium concentration to precipitation
Stage 6   second eluant to (c) to become first eluant in next cycle
Stage 7   as for stage 6 to completely elute (c)
Stage 8   column (c) on stand-by to become column (b) in next cycle
PL = pregnant liquor; W = water; 1 = 1st eluant; 2 = 2nd eluant; P = to precipitation; BD = barren discharge
(Reproduced by permission from 'Ion Exchange In Uranium Extraction', in 'Ion Exchange Sorption Processes In Hydrometallurgy', Critical Reports on Applied Chemistry, Vol. 19, ed. M. Streat and D. Naden, Wiley, Chichester, 1987, Ch. 1)

blind the resin beds, which presupposes a large and costly clarification–filtration plant to pretreat the leach liquor.

The uranium industry more than any other founded the commercial realization of Continuous Countercurrent Ion Exchange (CIX) technology which has resulted in being able to treat unclarified leach liquors in a near ideal continuous manner. Several modern CIX plants in the uranium industry are based on the successful Multistage

Fluidized Bed Contactor design pioneered by Cloete and Streat in the 1960s. The scale up of such a design is readily achieved to give very high flows of several hundred cubic metres per hour on a feed containing in excess of 1000 mg l^{-1} suspended solids. Upflow of solution in the loading and elution columns contacts resin maintained in a fluidized condition (bed porosity ~ 0.7) on specially designed perforated trays. During a brief period of reverse flow, resin is able to transfer stagewise from top to bottom of the columns whilst fully loaded and eluted resin is admitted to the top of the elution and loading columns respectively via resin transfer vessels. A simplified flow diagram is shown in Figure 9.3.

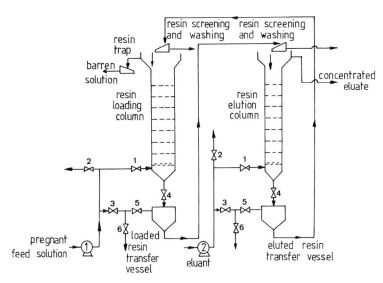

Figure 9.3 *Continous countercurrent fluidized bed ion exchange for uranium recovery. (1) Loading and elution column solution forward flow to fluidize the resin in each column, valve 1 open, valves 2, 3, 4, 5, and 6 closed. (2) No solution flow, resin settling, valve 2 open to return feed solution to storage, valve 1 closed. (3) Solution back flow, resin from bottom stage transferred to resin transfer vessels, valves 2, 3, 4, 5, and 6 open, valves 1 and 3 closed. (4) Valve flush and resin transfer from resin transfer vessels to top of fluid bed columns, no forward solution flow, valves 1, 3, 4, and 5 open, valves 2 and 6 closed. (5) Resumption of solution forward flow to the fluid bed columns.*

(Reproduced by permission from 'Ion Exchange In Uranium Extraction', in 'Ion Exchange Sorption Processes In Hydrometallurgy', Critical Reports on Applied Chemistry, Vol. 19, ed. M. Streat and D. Naden, Wiley, Chichester, 1987, Ch. 1)

Specific loading flowrates ($m^3\ h^{-1}\ m^{-3}{}_R$) are lower than for water treatment which reflects the particle diffusion contribution to the overall loading kinetics. Macroporous resins are preferred so as to minimize resin losses due to attrition, which nowadays may only amount to less than 10% of the resin inventory per year. Unclarified leach liquors may also be treated by contacting the resin and leach liquor in stirred tank contactors during the loading operation and then transferring the loaded resin to an elution station. Variations of this Resin In Pulp (RIP) process have been long established and are still under active development.

Some uranium ore bodies are high in limestone and dolomitic content which is wasteful on acid employed for leaching. Such sources are successfully treated by an alkaline mixed sodium carbonate and bicarbonate leach, the later being required to buffer the pH so as not to precipitate the uranium. Sorption occurs as the uranyl carbonate complex anion, $UO_2(CO_3)_3{}^{4-}$, but elution is with sodium nitrate since acid would cause evolution of carbon dioxide gas.

The economical feasibility of recovering heavy metals by ion exchange and liquid–liquid extraction is always under review. Such processes become increasingly appropriate as the unit price of an element rises, and the economics of extraction may be influenced favourably by the nature of new sources of the raw material. There has been much activity in the field of gold extraction by ion exchange and in the extraction of various elements from mine water, waste streams, dump leaching liquors, sea water, and other resources. Mostly strong base anion exchange resins are used but chelating resins play an increasing role. Liquid–liquid extraction techniques are generally preferable in those cases where a moderately concentrated liquor exists, but ion exchange operations remain more suited for treating dilute liquors or pulps and are generally more economical. Whatever the future developments in ion exchange extraction metallugy, the technology is now well advanced and many schemes are under active consideration.

BOX 9.4 Concentration of a Metal from a Dilute Solution

Sorption as a Simple Ion

Cation and anion exchange resins can be used as a means for obtaining a concentrated solution of a metal salt from a dilute solution. In this experiment, the principle is illustrated by loading a dilute solution of a nickel salt on to a SAC resin and eluting the nickel with a small volume of hydrochloric acid.

Non-water Treatment Practices

Requirements

Glass column, 12 mm diameter (approx.)
10 ml of SAC resin (H form)
500 ml of solution containing 500 parts per million of nickel (dissolve 1.25 g of $NiSO_4$ in 500 ml water or 1.40 g of $NiCl_2.6H_2O$ in 500 ml water)
20 ml of 5 N hydrochloric acid
10 ml measuring cylinder

Procedure

1. Put the resin into the column and pass the nickel solution through the resin bed at the rate of 2 BV/minute.
2. Add the 20 ml of 5 N hydrochloric acid and elute the nickel salt at the rate of 1 BV every 4 minutes.
3. Follow the acid elution with a 20 ml water rinse.
4. Collect the eluate in about eight portions of 5 ml each. The difference in the colour intensity of the fractions indicates the degree of concentration that has taken place.
   Alternatively, the nickel content in each fraction can be determined gravimetrically and an elution curve obtained by plotting a graph of volume of eluate against nickel concentration.

In this experiment the dilute nickel solution is concentrated about twenty times.

TIME REQUIRED – 1 hour 15 minutes

Sorption as a Complex Ion

This differs from the previous experiment in that the complex is formed with the cation already on the resin. A cation resin in the ammonium form is required. A solution of copper sulfate is used and the copper is absorbed on the resin as the blue tetramine complex.

Requirements

Glass column, 12 mm diameter (approx.)
5 ml of SAC resin (Na form)
5 N NH_4OH solution
1 litre of copper sulfate solution containing 4 ppm $CuSO_4.5H_2O$ acidified to pH 3 with sulfuric acid

Procedure

1. Put resin into the column and pass 5 ml of 5 N NH_4OH solution through the resin in 15 minutes. Rinse with 40 to 50 ml of deionized water as quickly as possible. The resin is now in the ammonium form.

2. Pass the copper sulfate solution through the column at 20 ml per hour. The copper is absorbed as the tetramine. This is a particularly slow experiment and the copper band on the resin is only seen after prolonged running.
3. The copper can be eluted with deionized water.

TIME REQUIRED – several hours

PHARMACEUTICAL PROCESSING

Not only is ion exchange a means of producing 'ultrapure' water for pharmaceutical preparations, but is widely adopted to concentrate and recover antibiotics and vitamins from fermentation broths.

For example, the moderately basic streptomycin molecule is very favourably exchanged by the sodium form of macroporous weak acid cation exchange resins. Elution is with dilute sulfuric or hydrochloric acid which efficiently strips the streptomycin to give a product of around 75% purity. Dilute sodium hydroxide solution is used to return the resin column to the sodium form. Any inorganic impurities are eliminated by conventional two-stage SAC–WBA demineralization using a *highly* crosslinked gel cation resin (whereby the large streptomycin cation is excluded on account of its size), and a gel weak base resin in the free base form. Organic colour bodies are removed by exchange across a strong base anion resin operated on the chloride cycle.

Given the monetary value of antibiotic agencies the ion exchange process usually operates on the triple column 'merry go round' design similar to that described previously for uranium recovery (Figure 9.2) since any potentially expensive losses or leakage are scavenged. Equilibration between the weak acid resin and antibiotic agent is so favourable that columns are able to operate upflow with the resin column in a partially fluidized (expanded) state thus enabling unclarified broth to be treated directly.

Clearly, the affinity of such large organic species for the resin is presumably enhanced by the complex matrix sorption mechanisms discussed under Chapter 5 ('Dilute Solution Cation Exchange'). Not only are ion exchange resins employed successfully for the recovery of the organic and biologically active agencies shown in Table 9.2, but the *non-functional* macroporous polymeric sorbents (Chapter 3, 'Solvent Modified Resin Adsorbents') very often demonstrate greatly superior performance. The latter are being used increasingly to recover biologically active agents including vitamin B_{12} and cephalosporin C where ion exchange resins may have been used previously.

Table 9.2 *Some biologically active substances recovered and purified by ion exchange*

Category	Examples
Antibiotics	Streptomycin; Neomycin; Cephalosporin C
Vitamins	B_1; B_6; B_{12}; C
Nucleotides	Adenylic acid; Cytidylic acid
Amino acids	Lysine; Glutamic acid; Methionine
Proteins and Enzymes	Albumin; Amylase; Deoxyribonuclease; Insulin
Viruses	Influenza; Mumps; Herpes

ION SEPARATION

Probably the best generally known separation process is that of chromatography where a mixture of species to be separated (mobile phase) is taken up by a sorbent (stationary phase) and then selectively desorbed in a manner governed by the relative sorption affinities of the various species for the sorbent.

Ion Exchange Chromatography

Here, the species to be separated are ions in solution (mobile phase) and the sorbent or stationary phase is a column of appropriately selected ion exchange resin. Hence ion exchange chromatography is a technique by which mixtures of ions in solution may be separated. Samuelson has presented a thorough survey of practical examples covering a host of elements.

The separation of ions by ion exchange chromatography often involves adjusting the distribution of ions between the exchanger and solution, especially if the ions have no significant selectivity over others for a particular resin. Some elements form charged complexes of a particular sign and separations may be based on a charge reversal principle with a particular resin. Kraus and co-workers have studied systems on this basis and have reported on the sorption characteristics of many ions in various media. In the majority of ion exchange chromatographic procedures, the ions to be separated are sorbed as a narrow band at the top of a resin column. A separation follows by using one of several displacement or elution techniques.

Consider a column of strongly functional cation exchange resin in the A^+ cation form upon which is loaded a small quantity of cations B^+, C^+, and D^+ given that the selectivity coefficient sequence is $K_A^D > K_A^C > K_A^B$. *Displacement development* employs a solution of another ion E which has the highest affinity for the resin ($K_D^E > 1$) which, as it passes down the column, displaces the counter-ions of the mixture with a self-sharpening boundary which in turn displace ion A^+ with a self-sharpening boundary. In time the mixture is resolved into discrete zones which emerge from the column, without an interval, in order of their increasing resin affinities. The chromatogram takes the form of Figure 9.4a with the effluent concentration remaining constant governed by the equivalent concentration of the displacing ion E.

Elution development on the other hand employs an ion of lowest affinity for the resin, commonly the same as the original ionic form of the resin A^+. Now ions migrate with non-sharpening boundaries at rates inversely related to their resin affinities in the presence of A^+ ions. Distinctly separated bands are formed which emerge from the column in order of their increasing resin affinities as depicted by Figure 9.4b. Complete resolution is possible by this method and is benefited kinetically by slow elution across resins of small particle size for a given system.

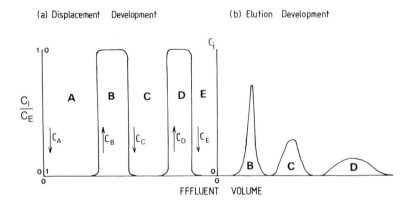

Figure 9.4 *Ion exchange chromatograms (C_i = equivalent concentration of ion i in the effluent; C_E = equivalent concentration of displacing ion E)*
 a) *Displacement development*
 b) *Elution development*

BOX 9.5 Elution Chromatography of Simple Cations

A mixture of cations may be separated by removing them from solution on a strong cation resin, and by eluting them off in the order of their increasing affinities for the resin. In this experiment, sodium, potassium, and magnesium are together loaded onto a SAC resin and removed separately with hydrochloric acid.

Requirements

Glass column, 12 mm diameter (approx.)
25 ml of SAC resin
0.1 M NaCl/0.1 M KCl/0.1 M $MgCl_2$ solution
Hydrochloric acid, 5 N, 2 N, 1 N, and 0.6 N.
Flame photometer and other means for analysis of the metals

Procedure

1. Put the SAC resin into the column. Regenerate it with 3 BV of 2 N HCl followed by 2 BV of 5 N HCl. Rinse until acid free with deionized water and drain to 6 mm above bed level. To remove trace metals and ensure that the resin is fully in the hydrogen form, a large excess of regenerant is used.
2. Pipette 3 ml of the Na, K, and Mg solution on to the top of the resin and slowly drain the liquid to the top of the bed.
3. Pass 0.6 N HCl at a rate of 1 BV/10 minutes through the resin and collect the eluate in 1 BV fractions. Analyse the fractions for cations. Typical results show that sodium starts eluting after 2 BV have been collected and ceases after 7 BV. Potassium elution starts at 8 BV and finishes after 14 BV have been collected. To speed up the elution of magnesium, the molarity of the hydrochloric acid is raised to 1 N. Sodium and potassium may be conveniently estimated using the flame photometer, and magnesium by EDTA (ethylenediaminetetraacetic acid).

N.B. a) The separation of Na, K, and Mg is improved at lower elution flow rates.
b) The technique of increasing elutant concentration to effect rapid separation is known as *gradient elution*.

TIME REQUIRED – 3 hours 30 minutes

Ions of very similar chemical characteristics have little difference in exchange affinity and therefore conventional chromatography techniques fail to give a good separation. However, in some cases, complexing agents may be used in the elution step to control the chemical activity of the various ions with respect to the resin and solution phases and, where there are large differences in the stability constants for the complexes, good separations are frequently possible.

The use of complexing agents for the elution of chemically similar ions is extremely important in the separation of the rare earths and post-uranium elements. The classical work and subsequent studies concerning the separation of the rare earths were conducted by Spedding, Ketelle, Boyd, and Tomkins who used buffered citric acid solutions as the elutant for loaded resins in the ammonium form. Further studies have demonstrated the usefulness of other complexing agents such as hydroxycarboxylic acids and EDTA (ethylenediaminetetraacetic acid). The role of ion exchange in isolating rare earth elements, post-uranium elements, and amino acids, so enabling their fundamental chemistry to be studied as well as their preparation in commercial quantities, is one of its greatest achievements.

High Pressure Ion Chromatography (HPIC)

High Pressure Ion Chromatography is a relatively recent (1975) and immensely significant development of conventional ion exchange chromatography by Small and co-workers of The Dow Chemical Company. This technique has revolutionized instrumental wet chemical analysis in that micro-amounts of an ion or mixtures of ions may be separated and assessed quantitatively in a matter of minutes compared with hours using traditional techniques. Accurate volumetric dispensing methods means that a quantitative analysis of ions may be achieved whatever their initial concentration.

The key components are shown in Figure 9.5 and comprise a high pressure pump, separator column, suppressor column, and a recording conductivity instrument.

Separator Column

The separator column contains a packing of *pellicular* resin (Chapter 2, 'Special Ion Exchange Materials') which owing to their small particle size (10–25 μm) present a high surface area and short diffusion paths therefore promoting very fast rates of exchange and rapid resolution of mixed ions. The resins used are of low capacity (0.005–0.02 meq g^{-1}) which for cation analysis are surface sulfonated copolymer microspheres. An anion exchange resin is obtained by surface coating a pellicular cation exchanger with a quaternary ammonium latex (0.1–1.0 μm particles) copolymer which becomes irreversibly electrostatically bonded to the host particle. Because of the deliberate low column exchange capacity the elution times for total ion loadings of only a few micrograms are very short. The eluting agents are prescribed according to the known or anticipated ions present and are pumped at a controlled rate (0.06–0.6 l h^{-1}) down the column.

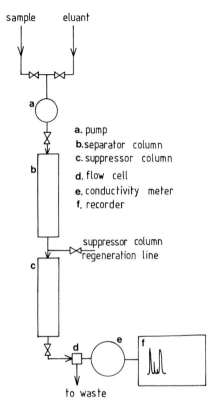

Figure 9.5 *Principal components of a HPIC system*

Suppressor Column

This is the heart of the technique together with the subsequent conductivity detection method, and is best explained by reference to some simple examples.

Cation Analysis (Figure 9.6a). Suppose a dilute acid HY is used to elute a cation mixture in the order B^+, C^+, D^+. The suppressor column contains a finely graded conventional anion exchange resin in, say, the hydroxide form. Prior to elution of ion B^+, the suppressor column sees only the eluant HY, and the following exchange takes place:

$$ROH + HY \rightarrow RY + H_2O \qquad (9.6)$$

Thus the conductivity of the column effluent is that of water and

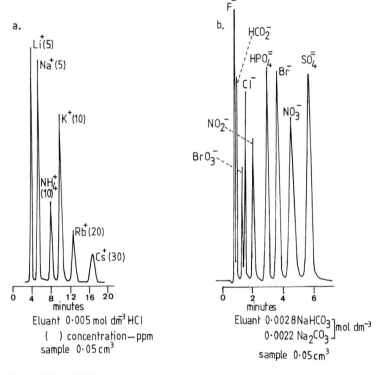

Figure 9.6 *HPIC ion chromatograms*
   a) *Cation analysis*
   b) *Anion analysis*

(Reproduced by permission from J. Weiss, 'Handbook of Ion Chromatography', ed. E. Johnson, Dionex Corporation, 1986)

therefore very low (*i.e.* has been suppressed). During elution of ion B^+ the suppressor column sees electrolytes BY and HY. The conductivity due to HY is nullified as described above, but BY exchanges to give highly conductive hydroxide ions:

$$ROH + BY \rightarrow RY + B^+ + OH^- \qquad (9.7)$$

When elution of ion B^+ is complete, only the eluant is present and the conductivity falls again to that of water, only to dramatically increase again as ion C^+ is eluted and lastly ion D^+. Thus ion detection is through a series of conductivity peaks whose location (retention time) and size (height and area) is characteristic of the ion present and its concentration respectively.

Anion Analysis (Figure 9.6b). For anion analysis the analogous principles apply, only now the eluant is, for example, the hydroxide or mixed carbonate/bicarbonate ion and the suppressor column a cation exchange resin in the hydrogen form, thereby suppressing conductivity by the formation of water and dissolved carbon dioxide respectively.

Advances in the HPIC technique have resulted in various other choices of eluant depending upon the particular system, and modern suppressor columns may use a packed hollow fibre membrane design. The principles remain the same and in combination with an additional ion pre-concentration column, detection limits of a fraction of a microgram per litre (ppb) are achievable for a host of inorganic and organic ions.

Ion Exclusion

A separation technique of some importance is that of Ion Exclusion. A mixture of ionized and non-electrolyte solutes is passed down a column of resin, cation or anion, in the same ionic form as a chosen ion of the electrolyte. No ion exchange occurs and the electrolyte is excluded from the resin on the grounds of the Donnan theory (see Chapter 5, 'Sorption of Non-exchange Electrolyte and the Donnan Equilibrium'). The resin merely acts as a sorbent for the non-electrolyte. After equilibration between resin and non-electrolyte the column is eluted with water which rapidly displaces the electrolyte followed by the non-electrolyte. One of the major commercial applications of this technique is for the separation of ethane-1,2-diol (ethylene glycol) from brines. The same basic principles apply to the separation of strong from weak electrolytes. Ion retardation resins or 'snake-cage' amphoteric exchangers of cationic and anionic functionality achieve similar results except that the order of elution with pure solvent is reversed. This type of separation is useful where the non-electrolyte is a macromolecule too large to penetrate a crosslinked resin.

FURTHER READING

'Ion Exchangers', ed. K. Dorfner, Walter de Gruyter, Berlin and New York, 1991.

R. M. Carlyle, 'Ion Exchange Processing of Petrochemicals', *Chem. Ind. (London)*, 1982, No. 16, 561.

'Ion Exchange Sorption Processes In Hydrometallurgy', Critical Reports on Applied Chemistry, Vol 19, ed. M. Streat and D. Naden, Wiley, Chichester, 1987.

'Recent Developments In Ion Exchange', ed. P. A. Williams and M. J. Hudson, Elsevier Applied Science, London and New York, 1990.

J. Melling and D. West, 'A Comparative Study Of Some Chelating Ion Exchange Resins For Applications In Hydrometallurgy', in 'Ion Exchange Technology', ed. D. Naden and M. Streat, Ellis Horwood, Chichester, 1984, p. 724.

'Ion Exchangers in Organic and Biochemistry', ed. C. Calmon and T. R. E. Kressman, Interscience, New York, 1957.

R. Kunin, 'Ion Exchange Technology In Medicine and The Pharmaceutical Industry', Amber-Hi-Lites, Rohm & Haas Co., No. 142–145, 1974–1975.

O. Samuelson, 'Ion Exchange Separations In Analytical Chemistry', Wiley, New York, 1973.

J. Weiss, 'Handbook Of Ion Chromatography', ed. E. Johnson, Dionex Corporation, 1986.

H. Small and J. Solc, 'Ion Chromatography – Principles and Applications', in 'The Theory and Practice of Ion Exchange', ed. M. Streat, Society of Chemical Industry, London, 1976, p. 32-1.

Chapter 10

Some Engineering Notes

CONVENTIONAL PLANT

The configuration of most ion exchange plants involve one or several fixed resin beds contained in cylindrical pressure vessels interconnected by pipework. Valves control and direct pumped or gravity fed liquid flows in a strictly controlled sequence, and safely designed regenerant measuring plus dilution systems are required to handle resin regeneration chemicals. Finally, instrumentation is required to monitor liquid levels, flows, pressures, volumes, and product quality.

Fixed Bed Coflow Units (Figure 10.1)

Vessel

For large columns over about 1.2 m diameter the usual material of construction is mild steel or sometimes stainless steel, with a height: diameter ratio of around 2:1 or greater in some cases. Access is via a top or side manway, and often the vessel carries a bottom spigot for resin removal. Windows, whilst not essential, are highly desirable to view resin levels, resin separations in mixed beds and layer beds, bed movement, and resin transfers. Most applications require the vessels to be pressure vessels and therefore are fabricated and tested to strict codes of practice.

Finally the chosen material of construction may be required to be internally lined to withstand the action of regenerant chemicals, process liquors, and possibly elevated temperatures. A specified rubber lining is still widely used, although glass or ceramic filled epoxy resins and plastic coatings are claiming wider acceptance. Small pressure vessels for domestic or laboratory use, and packaged industrial plant less than 1.2 m diameter of standard design are usually fabricated from an approved fibre reinforced plastic material.

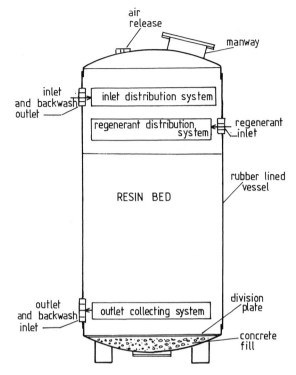

Figure 10.1 *Internal assembly of a coflow ion exchange vessel*

Internals

a) Inlet (Figure 10.2a): The inlet serves to admit feed or rinse water, and is also the backwash outlet. Designs are varied, but four types are quite common: simple bell mouth, header and laterals covered with a coarse mesh, header and laterals covered with a fine mesh, or an inlet strainer. Suitably covered laterals or strainers are the most common types often designed so as to prevent accidental resin loss on backwash.

b) Regenerant Distribution (Figure 10.2b): Usually a simple spreader plate or plain header and perforated laterals design to distribute regenerant across the whole bed area.

c) Outlet Collecting Systems (Figure 10.2c): Although simple in concept, the design of collection systems is critical so as not to allow pockets of regenerant to remain in the resin bed (hideout), and to exhibit a

Some Engineering Notes 263

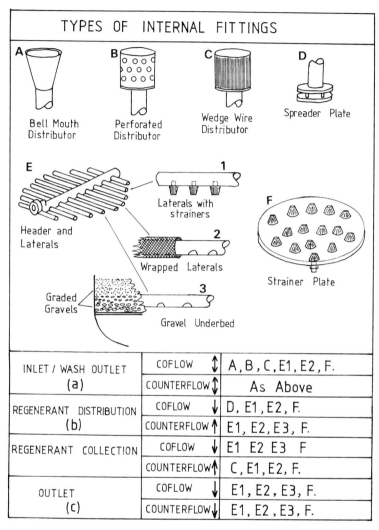

Figure 10.2 *Some common designs of fixed bed vessel internals*

suitable resistance to flow thereby ensuring even plug flow down the bed and collection across the entire bed cross-section. Also the outlet system has to accommodate variations in flow as governed by the dictates of service, backwash where applicable, rinses, and sometimes as an inlet for air (*e.g.* mixed bed units). Modern designs often use a collection header fitted with wrapped laterals or laterals carrying strainer nozzles with a bottom division plate supported by a concrete fill of the bottom dishing. Alternatively, the bottom outlet system could be a suitably supported division plate carrying a sufficient number of strainer nozzles discharging into a hollow dishing. Finally, one may still encounter a simple header and perforated laterals design, the latter being covered to a depth of 30–40 cm with a layer of graded gravels to support the resin bed.

Mixed Beds (see Chapter 8, 'Combined Cycle Single Stage Demineralization')

In situ regenerated mixed beds are of a typical design illustrated by Figure 10.3. A collector designed to be located at the separation interface of the cation and anion exchange resins serves as the anion resin regenerant/rinse collector and the cation resin regenerant/rinse distributor. The interface collector is commonly of the header and laterals type using a mesh wrapping or strainer nozzles. The outlet collector may be similarly constructed or of strainer plate design, and is also the distributor for backwashing (resin separation) and air (resin mixing). Some polishing mixed bed designs allow for simultaneous regeneration and displacement rinse of the resins where upflow acid (counterflow) and downflow caustic regenerants both exit at the interface collector. The technique saves on regeneration time, but

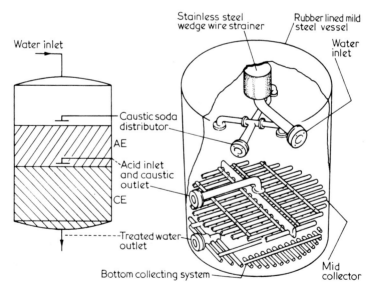

Figure 10.3 *Regenerant distribution–collection systems and internal arrangement for a typical mixed bed unit. (AE = anion exchanger; CE = cation exchanger)* (Reproduced from R. W. Grimshaw and C. E. Harland, 'Ion-exchange: Introduction to Theory and Practice', The Chemical Society, London, 1975)

cannot be used for hard water treatment working mixed beds since the contact of cation resin and anion resin regeneration effluents may cause precipitation of calcium and magnesium compounds within the bed and around the centre collector.

Counterflow Units (see Chapters 7 and 8)

The broad overall principles of vessel construction remain the same as for coflow units except for one crucial, and essential, exception. The counterflow mode of operation means that a conventional fixed bed will be unstable during regeneration or service depending upon whether employing upflow regeneration or upflow service respectively. Therefore all counterflow designs incorporate a means whereby the resin bed remains stable and packed at all times. Several 'hold down' techniques may be encountered in installed counterflow plants as illustrated in Figure 10.4.

a) Water Hold Down: This approach uses a downflow of water to keep the resin bed from fluidizing during upflow injection of regenerant and upflow rinse. Both the up and down flows discharge through a sub-resin surface collector placed about 30 cm below the top of the resin. The buried collector serves also as the backwash inlet for washing the topmost layer of resin without disturbing the counterflow regeneration portion of the column. The technique was one of the first used in counterflow designs, but is not highly favoured nowadays because of its wasteful regeneration water consumption.

b) Split Flow Regeneration: This technique is similar to water 'hold down' but uses actual regenerant solution instead of water. The simultaneous downflow and upflow regenerant discharges via a buried collector positioned about $\frac{1}{4}$ to $\frac{1}{3}$ of the resin depth below the bed surface. Thus the top portion of the column is regenerated coflow whilst the lower greater portion of the resin is regenerated counterflow. Not only does this approach allow for a more substantial sub-surface backwash, but also lends itself to arranging for different regenerant concentrations upflow and downflow. This may be particularly beneficial for stratified or layer bed systems incorporating two resins of weak and strong functionality.

c) Air Hold Down: Air 'hold down' is a proven and common full bed counterflow regeneration method. Here, a sub-surface (top) collector

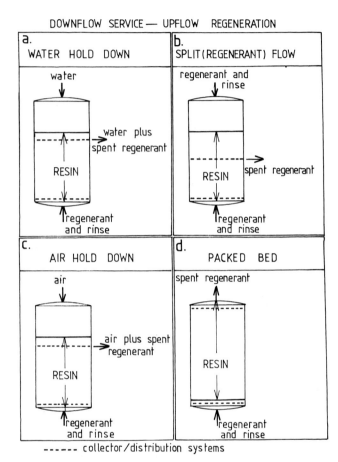

Figure 10.4 *Some common counterflow 'hold down' systems*

is covered by about 30 cm of resin or in some instances inert copolymer material is used of density less than the resin so as always to lie above the top collector. The column is drained to the level of the top collector and low pressure air admitted from the top of the vessel discharging at the collector. At the same time regenerant passes upflow through the compact resin bed and out to drain through the top collector. The air flow causes the resin or inert material above the collector to dry out giving a bridging effect, the effect of which maintains the resin bed in consolidated stable state. Again a partial backwash is possible via the sub-surface collector. Both single and stratified resin beds lend themselves to this particular counterflow regeneration technique.

d) Packed Bed: The fundametal requirement of fixed bed counterflow regeneration is that during regeneration and rinse the resin bed should not move. What better way to achieve this than by simply packing the vessel completely with resin leaving no freeboard at all in which the resin can move. In principle it is then possible to operate upflow regeneration and downflow service or *vice versa* whichever suits. This approach has become very popular in recent years because of its simplicity, and particular suitability to packaged and standard design. Of course, a small freeboard must be allowed to accommodate any initial irreversible and subsequent reversible volume changes between regenerated and 'exhausted' states of the resin. In fact a small voidage is always created initially since usually resins shrink when contacted by regenerant solutions.

If no provision is made for creating a backwashing facility then a packed bed design cannot tolerate suspended solids in the feed. Some designs can accommodate a light solids burden simply by transferring a part of the resin to a separate backwashing vessel. Alternatively specially designed compartmentalized vessels may be used in which backwashing space is created by the transfer of resin from one compartment to another.

Regenerant Handling

a) Measurement: Volumetric measurement of bulk stored liquor is most common using gravity head, vacuum eductors, or chemical pumps to charge a measuring vessel with concentrated regenerant. Some applications may require preparing a predetermined volume of regenerant solution from the solid chemical.

b) Dilution: This is commonly achieved by batch dilution of the previously measured charge of regenerant liquor which is then educted or pumped to the vessel. Some systems use a method whereby further interstage dilution of the regenerant and subsequent delivery to the vessel take place simultaneously.

Alternatively, where the chemical properties of the regenerant allow, the concentrated bulk regenerant may be diluted directly in-line using a pumped or eductor system. Whatever system is used safety of chemical handling is paramount, and the ideal design will incorporate interlocks and alarms which stop the delivery of bulk chemical should there be dangerous excursions in chemical concentration for whatever reason.

External Pipework

Commonly, mild steel, lined mild steel, stainless steel, or plastics are used. The actual choice depends upon the application, required structural strength, and relative costs.

Valves

Manual and automatic valves of the diaphragm, butterfly, or ball type are commonly used of a construction suited to the application. Automatic valves may be actuated pneumatically, hydraulically, or electrically and where large engineered plants are concerned separate valves are ascribed to discrete functions. Smaller packaged units often employ a single multiport valve mounted on the top or the side of the vessel.

Regeneration Sequence Control

Older generation plant employed motorized sequence controllers based on cams, timers, relays, and microswitches respectfully referred to by those in the field as 'rock and roll' gear. Nowadays microelectronics has taken over to provide programmable logic control (PLC) ranging from relatively inexpensive units associated with standard design packaged plant, to very sophisticated systems incorporating full graphical display of instantaneous plant status, event data logging, and fault finding facilities.

Instrumentation

Instrumentation serves three main functions:

i) To stop and restart a particular operation, for example the service cycle or column regeneration. Integrating volume throughput meters and rate of flow measurement are essential for the satisfactory operation of ion exchange plant. Hydraulically, continuity of plant output relies upon maintaining adequate liquor levels in feed liquor storage tanks, treated liquor storage tanks, and possible interstage tanks. Usually, electrode probes, float switches, and ultrasonic level detection devices control the stop–start status of a plant in service through an interlock with the appropriate valves and pumps. Electrode probes and ultrasonics are also used to measure bulk liquid regenerant chemical volumes, whilst rate of flow indicators serve to establish correct regenerant dilution. Finally in-line instrumentation to monitor product quality also serves to effect a shutdown should the quality in terms of the parameter measured fall below a preset value and to signal start-up when quality is restored again. Not all instrumentation need necessarily exercise a control function but instead may serve an indication purpose only, for example inlet and outlet pressure indicators across an ion exchange column.

ii) A second important function of plant instrumentation is to raise an alarm should a control limit or fault condition arise. It is not always desirable that an alarm signal should initiate a change in plant status automatically, but instead serve to draw the attention of a plant attendant who then decides an appropriate action to be taken. For example, in the common case of a batch integrator or quality monitor alarm indicating that column regeneration is required it may be inconvenient for actual regeneration to take place. In such circumstances the 'exhausted' plant will stand until an automatic regeneration is initiated manually.

Regeneration is the most important step of the operational cycle of an ion exchange plant and for this reason it is often desirable for the sequence to be halted should something go wrong at any stage. For example, flow sensors may be used to inhibit regeneration should the flow of regenerant dilution water fall too low. Conductivity measuring equipment may be used to monitor the concentration of a particular regenerant and also to inhibit a regeneration should the concentration deviate outside set limits.

iii) Finally, instrumentation is required to monitor the quality of the treated product not only at the final outlet of a plant but also at certain interstage locations depending upon the particular application. Conductivity measurement is widely adopted since its value is directly related to ion concentration. In principle, any suitable in-line sampling and measuring instrument may be employed providing it is robust, reliable, and measures a process related parameter with sufficient accuracy. For example, conductivity measurement hardness monitors (colorimetric), silica monitors (colorimetric), chloride monitors (electrode), sodium monitors (electrode), UV absorbance, colour, specific ion electrodes, spectrophotometry, and pH are just a few specific and general instrumental analysis techniques which find application in monitoring the on-line performance of ion exchange processes.

CONTINUOUS COUNTERCURRENT ION EXCHANGE

Conventional fixed bed ion exchange whether operated coflow or counterflow utilizes resin bed depths ranging from at least 0.75 m up to several metres. Exchange occurs within the migrating exchange zone across which the concentration of ions entering the column is reduced to zero. To achieve a practical net treatment rate ($m^3 h^{-1}$) it is essential that the total exchange capacity of the column be sufficiently large to achieve a service run time that is considerably longer than the outage time during regeneration. This situation is represented by Figure 10.5a where, at a given instant, the working volume of resin is only a relatively small fraction of the whole bed, the rest being either in an 'exhausted' or regenerated condition. At breakthrough the bed is taken out of service to await regeneration and flow to service maintained by switching to a standby column. The size, and therefore capital cost, of conventional fixed bed plant is governed to a significant extent by the need for a large resin inventory and multiplicity of plant.

For a given application *Continuous Countercurrent Ion Exchange (CIX)* aims to approach the ideal situation represented by Figure 10.5b where a small volume of resin equal in column length to the exchange zone is continuously moving countercurrent to the appropriate liquor flow during the loading (service), regeneration, and rinse stages of the overall cycle. Hydrodynamic constraints and ion load variations result in resin volume transfers being greater than theoretical and to obtain high exchange efficiencies at low capital cost it is necessary to establish a sufficiently high resin:liquor residence time ratio within a column stage. At any one time the various loading, regeneration, rinsing, and washing cycles are occurring simultaneously in separate columns as depicted purely schematically in Figure 10.5d, or the entire cycle may be accommodated by the various stages taking place within different zones of a single column. As might be anticipated the countercurrent mode of operation offers the benefit of high product quality (low 'leakage') and good regenerant utilization efficiencies for exactly the same reasons as described for counterflow fixed bed units.

Unlike fixed bed designs where scale up data may be obtained semi-empirically, continuous countercurrent ion exchange plant requires model hydrodynamic data for both the liquid and resin phases as well as predetermined equilibrium and kinetic data for a chosen system. A continuous cycle becomes particularly attractive when required to treat more highly concentrated liquors or operate at high treatment flowrates.

For a detailed account of the theoretical and practical design of

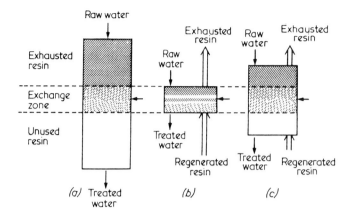

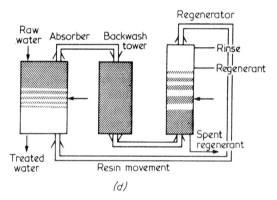

Figure 10.5 *A schematic comparison of fixed bed and continuous ion exchange*
  a) *fixed bed unit*
  b) *Ideal continuous absorber*
  c) *Practical continuous absorber*
  d) *Practical complete unit*
(Reproduced by permission from T. V. Arden, 'Water Purification By Ion Exchange', Butterworths, London, 1968, p. 127)

continuous countercurrent plant the reader is referred to the bibliography; but the most successful commercially proven designs are based on moving 'packed bed' systems such as the Higgins Loop contactor shown in Figure 10.6, and the ASAHI system or the Cloete–Streat multistage transfer fluidized bed concept, *e.g.* NIM-CIX. Continuous ion exchange technology is well established in the hydrometallurgical processing industry (uranium recovery) and has also been applied successfully in areas such as carbohydrate refining,

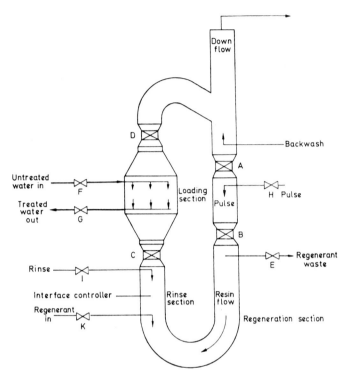

Figure 10.6 *The Higgins Loop continuous countercurrent ion exchange system*
Valve positions during cycles:
a) Run cycle: Valves A, E, F, G, I, K OPEN
   Valves B, C, D, H CLOSED
b) Pulse cycle: Valves B, C, D, H OPEN
   Valves A, E, F, G, I, K CLOSED

(Reproduced by permission from I. R. Higgins and R. C. Chopra, 'Chem-Seps Continuous Ion-Exchange Contactor and Its Application to Demineralisation Processes', in 'Ion Exchange In The Process Industries', Society of Chemical Industry, London, 1970, p. 121)

radioactive waste treatment, softening, demineralizing, and nitrate removal.

In the early days of development of continuous ion exchange processes the benefits to water treatment of high efficiency and low leakage were paralleled by the rapidly advancing designs of counterflow fixed bed plant. Thus the more complex hydrodynamic requirements of CIX, resin losses due to mechanical attrition, and perhaps a conservative attitude towards availability of plant during periods of unscheduled maintenance meant that, in the UK at least, the CIX technology did not gain the foothold in water treatment that might have been expected. CIX designs have improved greatly in all respects, but currently their niche would seem to lie largely outside the field of conventional water treatment.

THE PAST, PRESENT, AND FUTURE

Table 10.1 shows the chronological order of key developments in resin synthesis, process application, and plant design, not so much from a literal historical standpoint, but more a classification according to the principal decade during which significant advances occurred. Since the advent of organic polymeric ion exchange resins it is not surprising that technological advances and resin developments should quickly follow each other since both impact on obtaining optimum plant performance. The summary given in Table 10.1 begs the question as to what of future developments?

Ion Exchange Resins

Modern organic ion exchangers are stable high quality materials which go a long way towards meeting the ideal properties listed at the beginning of Chapter 2. From an ion exchange point of view development is on-going to investigate different copolymers and functional groups with a view to enhancing specific ion selectivity, for example chelating resins; or to modify ion selectivity sequences as was demonstrated in the case of nitrate removal. New applications will certainly arise as supported by recent reports describing the bacteria removing properties of a weakly crosslinked anion exchanger (4-vinylpyridinium bromide) deposited within a porous absorbent such as silica gel or a zeolite, *i.e.* an insoluble disinfectant.

Where the physical properties of resins are concerned manufacturers continue to develop products offering maximum physical stability, and more recently, of uniform bead size.

Engineering

Whilst it is not essential to comment on advances in detailed equipment design there are, nevertheless, changing trends taking place which might be best described as a strive towards 'low capital cost engineering'. A large part of the capital cost of any plant is the cost of

Table 10.1 *Key developments in resins, engineering, and processes*

Period	Developments in Ion Exchange and Related Technology
1940	Styrene–DVB resins (cation and anion) First ion-specific resin synthesized Early industrial 2-stage demineralizing First commercial mixed bed demineralizing
1950	Theoretical advance (equilibrium, kinetics, mechanism) Acrylic weak acid cation resins Commercial uranium extraction by ion exchange First UK nuclear power station – Calder Hall Classical chromatography
1960	Theoretical advances ongoing Continuous countercurrent practice and development Fixed bed counterflow designs Macroporous, Isoporous, and Acrylic anion resins 'The Organic Fouling Debate'
1970	Ion exchange desalination – 'DESAL', 'SIROTHERM' Membrane desalination – Reverse Osmosis, Electrodialysis Continuing fixed bed counterflow development Condensate polishing systems 'Ion Chromatography' analysis, and Pellicular resins Polymeric adsorbents
1980	Advanced Condensate Polishing designs 'Sub-ppb' ion residuals established for condensate treatments and 'Ultrapure' water applications Nitrate selective resins Resin copolymerization – uniform bead size
1990	Commercial nitrate removal – UK High purity resins – 'low leachables' 'Packaged Engineering', and standard designs Trailer mounted non-regenerable plant Short Cycle ion exchange technology Advances in techniques for TOC removal, *e.g.* activated irradiation (UV), ultrafiltration

design, and manufacture of components. Therefore it is a cost benefit to adopt standard designs of component items wherever possible. Ultimately the capital cost of a plant is governed by the net production output required and the raw feed ion concentration. For simplicity, consider a single ion exchange unit requiring to use treated product in its regeneration cycle. The net output, F (m^3 h^{-1}) from the unit is given by the expression:

$$F = \frac{\frac{V_R C_R}{L} - W}{T + R}$$

where

V_R = Resin volume (m^3)
C_R = Resin operating capacity (keq m$^{-3}_R$)
W = Treated product volume used during regeneration (m^3)
T = Column run time (h)
R = Regeneration time (h)
L = Ionic concentration (keq m^{-3})

For a given fixed bed design and ionic load the net output will benefit from a low regeneration volume W and a short cycle time ($T + R$). Very short cycle counterflow ion exchange plants are now established where run times are typically 1–2 hours with a regeneration time of 30 minutes or can be as low as a 15 minute run time and 5 minute regeneration period. Obviously such an approach dramatically reduces the size of plant required for a specified output on a given feed. A short cycle plant design, through its reduced plant size, lends itself to standardized system engineering and is proven commercially for water demineralization in the power, electronics, and pharmaceutical industries. In fact, fixed bed counterflow short cycle technology may be applied to ion exchange generally and if taken to the extreme could be seen as near continuous 'fixed bed counterflow ion exchange'.

The ionic concentration to be treated is an overriding consideration governing the cost of plant of a given design and therefore for very high ion concentrations it is foreseeable that membrane pretreatments such as reverse osmosis and electrodialysis will continue to fulfil an important role as might a more widespread revival of continuous countercurrent ion exchange.

Finally, attitudes are changing in the UK as to how the process industries perceive ownership of ion exchange plant especially for the provision of essential services, such as steam, power, and process water. Serious thought is being given to purchasing services from a plant built, owned, maintained, and operated by others either on or off site. This concept is already well established in the privatized power industry and to extend this practice to include process steam and water is a commercial consideration only. Traditionally, industrial ion exchange water treatment plants are permanently located and equipped with their own dedicated regeneration and effluent handling facilities.

Conversely, small laboratory units providing deionized water are kept operational by an exhausted resin cartridge being exchanged for regenerated resin through an off-site regeneration–cartridge exchange delivery service. However, the latter scheme is no longer confined to the small water user. Instead of resin cartridges read large pressure vessels containing sufficient regenerated resin to operate for 8–12 hours. Instead of a postal type exchange service read trailer-mounted containers housing plant for immediate 'hook-up' on site whilst the 'exhausted trailer package' is transported off-site for regeneration. Off-site regenerated ion exchange plant of the type described above and capable of outputs of several hundred cubic metres per hour are operating commercially. Whether such an operation is confined to providing water to cover for breakdown and major overhaul of permanent plant or as an alternative to owning plant is, again, a commercial rather than technical consideration.

FURTHER READING

Panel Discussion, in 'Ion Exchange In the Process Industries', Society of Chemical Industry, London, 1970, p. 143.

F. L. D. Cloete, 'Comparative Engineering And Process Features Of Operating Continuous Ion Exchange Plants in Southern Africa', in 'Ion Exchange Technology', ed. D. Naden and M. Streat, Ellis Horwood, Chichester, 1984, p. 661.

T. V. Arden, 'Raw Water Treatment By Ion Exchange', in 'Ion Exchangers', ed. K. Dorfner, Walter de Gruyter, Berlin and New York, 1991, p. 717.

M. Streat, 'General Ion Exchange Technology', in 'Ion Exchangers', ed. K. Dorfner, Walter de Gruyter, Berlin and New York, 1991, p. 685.

T. V. Arden, 'Water Purification By Ion Exchange', Butterworths, London, 1968.

M. J. Slater, 'Principles of Ion Exchange Technology', Butterworth–Heinemann, Oxford, 1991.

Appendix

USEFUL CONVERSIONS

Measure		Unit	← × ← / → ÷ →	Unit
Linear	Feet	ft	3.281	= m
	Angstrom	Å	10^{10}	= m
	Angstrom	Å	10	= nm
Area	Square feet	ft^2	10.764	= m^2
Volume	Cubic feet	ft^3	35.32	= m^3
	Imperial gallons	gall. (UK)	220	= m^3
	American gallons	gall. (US)	264	= m^3
Mass	Pound	lb	2.205	= kg
	Kilograin	Kgr	15.432	= kg
	Ounce	oz	35.27	= kg
Pressure	Pounds force per square foot	$lbf\,ft^{-2}$	0.0209	= Pa ($= N\,m^{-2}$)
	Pounds force per square inch	$lbf\,in^{-2}$	14.504	= Bar
	Atmospheric pressure	atm	0.987	= Bar
Work and Energy	Calorie	cal	0.239	= J ($= 10^7$ erg)
	British Thermal Unit	BTU	0.948	= kJ
	Calorific value	$BTU\,lb^{-1}$	0.430	= $J\,g^{-1}$
	Watt hour	Wh	0.2777	= kJ
Flow and Velocity		gall. (UK) min^{-1}	3.67	= $m^3\,h^{-1}$
		gall. (UK) $min^{-1}\,ft^{-2}$	0.34	= $m^3\,h^{-1}\,m^{-2}$ ($= m\,h^{-1}$)
Concentration and Density	Grams ($CaCO_3$) per litre	$g(CaCO_3)\,l^{-1}$	50	= $eq\,l^{-1}$
	Kilograins ($CaCO_3$) per cubic foot	$kgr(CaCO_3)\,ft^{-3}$	21.8	= $eq\,l^{-1}$
	Grains per gallon (UK)	$gr\,gall.\,(UK)^{-1}$	70.0	= $g\,l^{-1}$
		$gr\,ft^{-3}$	0.4356	= $mg\,l^{-1}$
		$lb\,ft^{-3}$	0.0624	= $kg\,m^{-3}$
	Grams per 100 cm^3 solution	% w/v	SG*	= % w/w

*SG = specific gravity of solution

SOME STANDARD RESINS ACCORDING TO GENERIC TYPE AND MANUFACTURER

a) Cation

Manufacturer Brand name	Type	Purolite Int. Purolite Purofine*	Rohm and Haas Amberlite Amberjet*	Duolite	Dow Chemical Dowex Monosphere*	Bayer Lewatit
Strong Acid Cation	gel gel macroporous macroporous	C100 C100 ×10 C150 C160	IR 120 IR 132 IR 252 IR 200	C225 C255 ⎱C26	HCR-S HGR-W2 — MSC-1	S 100 S 115 SP 112 SP 120
Weak Acid Cation	gel macroporous	C105 C106	IRC 86 IRC 50	C433 C436	CCR-2 MWC-1	— CNP 80
Chelating	iminodiacetate– macroporous aminophosphonic– macroporous	S930 S940	IRC 718 —	ES466 ES467	XZ 95843 XZ 87480	TP 208 OC 1060

*Unsized Product Range

b) Anion

Manufacturer	Purolite Int.	Rohm and Haas		Dow Chemical	Bayer
Brand name	Purolite Purofine*	Amberlite Amberjet*	Duolite	Dowex Monosphere*	Lewatit
Type					
Strong Base Type 1					
gel–styrene	A400	IRA 420	A113	SBR-P	M 504
gel–acrylic	A850	IRA 458	A132	—	—
macroporous–styrene	A500	IRA 900	A161	MSA-1	MP 500
macroporous–styrene	A500P	IRA 904	A171SC	11	MP 500A
macroporous–acrylic	A860	IRA 958	A173	—	—
Strong base Type 2					
gel–styrene	A200	IRA 416	A116	SAR	M 600
macroporous–styrene	A510	IRA 910	A162	MSA-2	MP 600
Mixed Base					
gel–acrylic	A870	IRA 478	A134	—	AP 246 (macro)
Weak Base					
gel–acrylic	A845	IRA 67	A375	—	—
macroporous–acrylic	A830	IRA 35	A374	—	AP 49
gel–styrene	A110	IRA 47	A30B	WGR-2	—
macroporous–styrene	A100	IRA 93	A378	MWA-1	MP 64
macroporous–styrene	A103	—	A368	—	MP 62

*Unisized Product Range

Subject Index

Entries in italics are people's names

Acrylic anion exchange resins, 30, 46
Acrylic cation exchange resins, 27
Actinides, vii
Activated carbon, 47
Activation energy, 137
Activity, 93
Activity coefficient, 93, 110, 112
Adams, 22, 24
Addition polymerization, 25
Adsorbents, polymeric, 47, 252
Affinity, 3, 90
Air mix, 217, 264
Air hold down, 265
Alkalinity, 183, 192
All volatile treatment, 222
Aluminium sulfate, 3, 10
Aluminosilicates, 3, 10
Amination, 29
Amino acids, vii, 256
Aminophosphonic group, 33, 191
Ammonia cycle, 222
Ammonium sulfate, 2
Amphoteric exchangers, 35
Analcite, 17
Anion exchange, 1, 127
  resins, 25, 28
Anodizing, 229
Antibiotics, 252
Antigorite, 8
Arden, 247
Argensinger, 111
Aristotle, 2
Asahi (CIX), 271
Asbestos, 8

Backwashing, 168, 178, 216
Bacteriological activity, 224
Bakelite, 23
Basal plane (spacing), 4, 13, 14
Base Exchange, 2

Bauman, 115
Bed dressing, 168
Benzoyl peroxide, 25
Benzyltriethylammonium group, 130
Benzyltrimethylammonium group, 29
Bifunctional anion exchange resins, 28, 32
Bifunctional cation exchange resins, 28
Biotite mica, 8
Birefringence, 42, 43
Blowdown, 195
Boiler, 195, 218
Boiling water reactor, 232, 234
Bonner, 111
Boron, 234
Boyd, 115, 118, 145, 226
Brackish waters, 227
Bragg, 3
Breakthrough capacity, 168
Breakthrough curve, 159, 161
Broken bonds, 10
Brucite, 4
Buried collector, 265

Calcium carbonate ($CaCO_3$), 182
Carbohydrate refining, 238
Carbonaceous exchangers, 22
Carboxylate group, 27, 53, 105, 190, 192
Carix process, 226
Cartridge, 276
Catalysis, 238, 241
Cation exchange, 1, 2, 123
  resins, 22, 25, 27
Cation polisher, 215
Caustic slip, 208
Cavities, 16
Cephalosporin C, 252
Chamosites, 8
Channelling, 169
Channels, 16

Subject Index

Characterization of ion exchange resins, 49
Chelates, 32, 138
Chelating resins, 32, 191, 250
Chemical kinetics, 141
Chemical rate control, 138
Chemical stability of resins, 69
Chlor-alkali industry, 190
Chlorites, 4, 8, 10
Chloromethylation, 28
Chromate, chromic acid, 229
Chrysotile, 8
Clay minerals, 3
Clayton, 159
Clifford, 129
Cloete, 249, 271
Cloete–Streat multistage contactor, 271
Coflow regeneration, 174, 205, 212
Co-ion, 1
Collecting systems, 217, 262, 264
Colloidal humus, 3
Column
  breakthrough, 173
  dynamics, 158, 173
  operations, 167, 173
Combined cycle demineralization, 215
Condensate polishing, 218
Condensation polymers, 22
Constant pattern, 161
Continuous electrodeionization (CDI), 228
Continuous ion exchange, 248, 270
Co-ordination, 4, 8, 15
Copper, 228
Copolymer matrix, 24
Corn starch, 238
Corn syrup, 239
Counterflow regeneration, 176, 213
Counterflow units, 265
Counter-ion, 1, 21
Covalent bonds, 4, 8
Cross contamination, 221
Crosslinked copolymer, 23
Crosslinking, 24, 25, 62

D'Alelio, 25
Dealkalization, 192
Demineralization, 204, 205, 212, 213, 215
Density, resin, 84, 85
Desalination, 226
Desal process, 226
Dextrose, 239
Dichromate, 229
Dickite, 8
Diethenylbenzene (divinylbenzene, DVB), 25
Diffusion, 135

Diffusion coefficient, 136
Diffusion kinetics, 145
Dimethylaminopropylamine (DMAPA), 30
Dimethylethanolamine, 29
Dioctahedral silicates, 8
Displacement, 175
Displacement development, 254
Distribution systems, 217, 264
Donnan
  equilibrium, 101, 210, 259
  potential, 101, 108
Double layer, 5
Double layer lattice silicates, 8, 13
Dow Chemical Company, 256
Drugs, vii
Dry weight capacity, 70

Economics, 172
Effective size, 82
Efficiency, regeneration, 169, 170
Effluent treatment, 228
Eichorn, 115
Eisenman, 124
Electrochemical potential, 93
Electrodeionization, 228
Electrodialysis, 34, 227
Electrolyte sorption, 92
Electroneutrality, 2
Electron exchangers, 35
Electronics, 224
Electroplating, 229
Electroselectivity, 108, 170, 188
Elution, 175
  development, 254
  leakage, 161, 175, 206, 224
Energetics of ion exchange, 116
Enthalpy (ΔH), 116
Entropy (ΔS), 116
Equilibria, 90
Equilibrium isotherm, 106
Equivalent, 2
Equivalent mineral acidity (EMA), 185, 210
Ethenylbenzene (vinylbenzene), 25
Ethylene dichloride, 42
European Community/Union, 180
Exchange isotherm, 106, 188
Exchange zone, 159, 270
Expanding layer lattice, 14
Exposed hydroxyl groups, 12

Faujasite, 17
Favourable equilibrium, 161
Felspar, 16
Fertility, vii
Fibrous ion exchangers, 37

Fick's Law, 136
Field strength, 125
Film, 135, 155
Film diffusion, 135, 149, 151
Fixed anion, 1
Fixed bed (coflow), 261
Fixed ion, 21
Finite volume, 147, 150
Flux, 136
Framework structures, 16
Franks, 125
Free mineral acidity (*FMA*), 206
Fructose, 238
Fulvic acid, 44, 47, 197
Functional group, 24, 62

Gans, 18
Gas cooled reactors, 232
Gel resin structure, 40, 41, 42, 45
Gibbs–Donnan equilibria, 93
Gibbs–Duhem equation, 112
Gibbs free energy (ΔG), 116
Gibbsite, 4
Glauconite, 36
Glucose, 238
Glueckauf, 115
Gold, 246
Gregor, 94, 123

Half time, reaction, 154
Halloysite, 8
Harned's rule, 115
Harris, 124
Heavy water reactor, 232
Helfferich, 108, 145, 148, 151, 154
Heteroporous, 40, 45
Higgins loop contactor, 271
High fructose corn syrups (HFCS), 239
High pressure ion chromatography (HPIC), 256
Hogfeldt, 111
Holmes, 22, 24
Humic acid, 44, 47, 197
Hydrocarbon matrix, 22
Hydrogen bonding, 7, 129
Hydrolysis
 of hydrogensulfate ion, 221
 of weak base sites, 221
Hydrophilic structure, 2, 21
Hydrophobic bonding, 126
Hydrophobic hydration, 125
Hydroxonium ion, 22

Igneous rock, 4
Illites, 15
Iminodiacetate group, 32, 191
Inert resin, 222

Infinite volume, 146, 150
Injection, 169
Inorganic ion exchange materials, 2
Insitu regeneration, 217
Instrumentation, 268
Interactions, 123, 124, 125, 128
Interdiffusion coefficient, 148
Interface collector–distributor, 217, 264
Inter-lamellar bonding, 14, 15
Inter-lamellar sites, 14, 15
Internals, 262
Interruption test, 157
Interstitial ions, 16
Inversion, sucrose, 238
Ion equivalent mass, 182
Ion exchange, 1, 10
 capacity, 3, 70
 chromatography, 167, 253
 equilibria, 90
 kinetics, 134
 resins, 21, 25
 resin structure, 39
Ion exclusion, 240, 259
Ionic bonds, 4, 15
Ionic form, 62
Ionogenic group, 24, 62
Ion removal, 166
Ion retardation, 35, 259
Irreversible swelling, 69
Isomorphous substitution, 10, 13, 14
Isopiestic method, 97
Isoporous resin, 40, 45
Isotherm, 106, 188
Isotopic exchange, 146, 150

Kandite group, 5
Kaolin, 5
Kaolinite, 8, 10
Katchalsky, 124
Ketelle, 256
Kielland, 112
Kinetic leakage, 141, 175, 221
Kinetics, 134
Kraus, 253

Langmuir, 90
Lanthanides, vii
Laterals, 262
Layer beds, 215
Layer lattice silicates, 4, 5, 8
Leach liquors, 246
Leakage, 161, 173, 175, 206, 224
Lemberg, 17
Leucite, 17
Ligand exchange, 34, 239
Limited bath, 147
Linear driving force, 149, 150, 151

Ling, 124
Liquid exchangers, 34
Liquid–liquid extraction, 231, 247, 250
Lithium, 234
Livesite, 8
Loading, column, 168

Mackie, 137
Macroporous resins, 18, 40, 41, 45, 46, 47
Magnox reactor, 232
Make-up plant, 195
Manganese 'zeolite', 36
Mass transfer, 135
Mass transfer coefficient, 151
Matrix, resin, 58
Matrix charge separation, 129
McBurney, 28
Mean ionic activity coefficient, 93
Meares, 137
Mechanical models, 123, 128
Mechanism criteria, 155
Medicine, vii
Membrane materials, 34
Membrane filtration, 225
'Merry Go Round' process, 247, 252
Metal recovery, 246
Metathesis, 166, 244
Metered capacity, 168
Methacrylic acid, 27, 54
Methyl tert-butyl ether (MTBE), 242
Methylene group bridging, 29, 44
Methylpropenoic acid, 27, 54
Mica, 4, 8, 14
Michaels, 161
Microporous resins, 40
Mixed beds, 215, 217, 218, 225
Mobile systems, 276
Molecular models, 123, 128
Molecular sieves, 17
Monomer, 25
Montmorillonite, 9, 14
Multicomponent equilibria, 112
Multi-stage fluidized beds, 249
Muscovite mica, 8
Myers, 115

Nacrite, 8
Nair, 197
Nernst film (or layer), 135
Nernst–Plank equation, 148
Nimcix (CIX), 271
Nitrate removal, 202
Nitrate selective resins, 130, 202
Non-conventional resins, 37
Nonsol method, 46
Non-water treatment, 238

Nuclear applications, 18, 231

Octahedral layer, 9
Old Testament, 2
Off-site regeneration, 276
Operating capacity, 169
Operator vessel, 222
Organic fouling, 44, 209
Organic removal, 197
Organic trap, 200
Osmotic activity, 94, 96
Osmotic coefficient, 115
Oxidation–reduction resins, 35
Oxygen absorbed (OA), 197
Oxygen removal, 203

Packed-bed, 267
Particle diffusion, 137, 145, 147
Particle size, 13, 82, 155
Pauley, 124
Pauling, 3
Pellicular resins, 37, 256
Pepper, 95
Permanent hardness, 183
Permutit, 17, 18, 22
Petrochemical cracking and reforming reactions, 17, 242
Pharmaceutical processing, 252
Phenol–formaldehyde ion exchangers, 23, 24
Physical specification of resins, 82
Pickling, 231
Pipework, 268
pK value, 50
Polymeric resin adsorbents, 47, 252
Polymerization, 22, 25
Porter, 247
Post-uranium elements, 256
Powdered resins, 37
ppb, ppm, 180
Prediction of selectivity, 113
Pressure drop, 168
Pressurized water reactor, 232, 234
Primary amine, 24
Process design, 163
Propenoic monomers, 27
Proportionate pattern, 162
Purification, 166
Pseudo-cubic structure, 17
Pyrophyllite, 8

Quadratic driving force, 149
Quaternary ammonium group, 29, 31
Quentin process, 238

Radioactive wastes, 231
Rare earths, 222

Rate equations, 140
Rate mechanism, 140
Rational thermodynamic selectivity, 111
Rational scale, 107
Reaction half time, 154
Recovery processes, 166, 246
Regenerant handling, 267
Regeneration, 168, 169
   cycle, 174, 176
   efficiency, 169, 170
   sequence control, 268
Reichenberg, 120, 124, 156
Relative affinity, 104
Resin fines, 169
Resin in pulp (RIP), 250
Resin transfer, 222
Reverse demineralization, 238
Reverse osmosis, 187, 225
Reversible swelling, 68
Rinse, 169

Salt splitting capacity, 79
Samuelson, 253
Secondary amine, 24
Selectivity, 3, 90, 113, 131
Selectivity coefficient, 105
Semiconductor, 224
Separation, 166, 253
Separation factor, 106
Separation–regenerator vessel, 222
Separator column, 256
Sequence control, 268
Serpentines, 4
Service cycle, 168
Shell progressive mechanism, 139
Short cycle ion exchange, 275
Shrinking core mechanism, 139
Silica, 207
   sol, 244
   tetrahedron $(SiO_4)^{4-}$, 4
Single layer lattice silicates, 5, 10
Sirotherm process, 226
Slip, 161, 175
Small, 256
Smectites, 9
Snake-cage polyelectrolytes, 35, 259
Sodium leakage, 208
Softening, 187, 190
Sol method, 45
Solvent modified adsorbents, 47, 252
Solvent modified ion exchange resins, 45
Sorption
   of non-electrolytes, 100
   of non-exchange electrolyte, 101
   of solvents, 93, 100
Special ion exchange materials, 32
Specific ion exchange resins, 32

Spedding, 256
Sphericity of ion exchange resins, 87
Split-elution, 247
Split flow system, 265
Standby plant, 168
Strain free resin, 44
Stratified beds, 215
Streat, 113, 197, 249, 271
Streptomycin, 252
Strong acid–base functionality, 50
Strong base Types 1 and 2, 29, 63
Strongly acid cation resins, 26, 105
Strong–weak capacity, 77
Structure of ion exchange resins, 39, 59
Styrene, 25
Styrenic ion exchange resins, 25, 28
Sucrose (sugar), 238
Sulfate–hydrogensulfate hydrolysis, 220
Sulfonated coal, 22
Sulfonic acid, 22, 26
Suppressor column, 256
Sweeten on/off, 168, 169
Swelling, 68, 69, 91, 93
Swelling pressure, 94, 96

Talc, 4, 8
Temporary hardness, 183, 187
Tertiary amine, 24
Tetra-alkylammonium functionality, 95
Tetrahedral layer, 8
Thermal shock, 81
Thermal stability, 79
Thermodynamics of equilibria, 114
Thompson, 2, 90
Thoroughfared regeneration, 213, 215
Tomkins, 256
Total alkalinity, 183, 192
Total anions/cations, 184, 185, 206, 207
Total hardness, 183
Total organic carbon (TOC), 197, 200
Trailer-mounted plant, 276
Transition metals, 32, 138
Trioctahedral silicates, 8
Types 1 and 2 anion exchange resins, 29, 105

Ultrafiltration (UF), 225
Ultrapure water, 224
Ultraviolet (UV) absorbance, 197
Ultraviolet (UV) sterilization, 225
Unfavourable equilibrium, 162
Uniformity coefficient, 82
Units (analysis), 180
Unlimited bath, 146
Uranium, 246
Uranyl carbonate ion, 250
Uranyl sulfate ion, 247

Valves, 268
van der Waals forces, 30, 47
van't Hoff equation, 116
Vermeulen, 149
Vermiculite, 10, 14
Vitamins, 244, 252
Vinylbenzene (ethenylbenzene), 25
Vinyl polymerization, 25

Water
  analysis, 180
  content, 65
  cooled reactors, 234
  demineralization, 204
  hold down, 265
  regain, 66
  softening, 187
  structure enforced ion pairing, 128
  treatment, 179
Way, 2, 90
Weak acid cation resins, 27, 53, 105
Weak base anion resins, 30, 54, 63
Weber, 129
Wet volume capacity, 76

Zeo-Karb, 22
Zeolite (A, X, Y), 17
Zeolites, 16
Zero solids treatment, 222
Zinc, 228

Also of interest from the Royal Society of Chemistry...

Ref No 779

Ion Exchange Processes: Advances and Applications

Edited by A. Dyer, *University of Salford*
M.J. Hudson, *University of Reading*
P.A. Williams, *The North East Wales Institute, Deeside*

This new book covers all aspects of industrial, analytical and preparative applications of ion exchange and presents topical reviews of subjects such as pharmaceutical preparation analysis, potable water treatment, the nuclear power industry, inorganic materials, the production of ultrapure water and the design of new chelating exchangers.

Ion Exchange Processes: Advances and Applications has an international authorship and is written by experts whose interests span the all-pervasive influences and applications of ion exchange. They have provided information on the latest advances in their fields, making this book essential reading for researchers from both industry and academia with involvement in this field.

Brief Contents:
Part 1: Fundamentals of Ion Exchange; Part 2: Ion Exchange in the Nuclear Industry; Part 3: Capillary Electrophoresis; Part 4: Water Treatment; Part 5: Inorganic Ion Exchangers; Part 6: New Materials; Subject Index.

Special Publication No. 122

Hardcover   x + 372 pages   ISBN 0 85186 445 7(1993)   Price £52.50

To order please contact:

Turpin Distribution Services Ltd., Blackhorse Road,
Letchworth, Herts SG6 1HN, UK.
Tel: +44 (0) 462 672555. Fax: +44 (0) 462 480947.
Telex: 825372 TURPIN G.

For further information please contact:

Sales and Promotion Department, Royal Society of Chemistry,
Thomas Graham House, Science Park,
Milton Road, Cambridge CB4 4WF, UK.
Tel: +44 (0) 223 420066. Fax: +44 (0) 223 423623.
E-mail: (Internet) RSC@RSC.ORG.

ROYAL SOCIETY OF CHEMISTRY

Information Services